QUANTITATIVE APTITUDE STATISTICS

QUANTITATIVE APTITUDE STATISTICS

Dr. Alok Kumar

RANDOM
RANDOM PUBLICATIONS
NEW DELHI - 110 002 (INDIA)

Quantitative Aptitude Statistics

ISBN 978-93-51117-11-7

Published in 2015 in India by

RANDOM PUBLICATIONS

4376-A/4B, Gali Murari Lal, Ansari Road
New Delhi-110 002
Phone: +9111-43580356, 23289044
E-mail: randomexports@gmail.com; sales@randompublications.com; info@randompublications.com

Reprinted 2026

Type Setting by: Friends Media, Delhi-110089

Preface

An aptitude is a component of a competency to do a certain kind of work at a certain level, which can also be considered "talent". Aptitudes may be physical or mental. Aptitude is not developed knowledge, understanding, learned or acquired abilities (skills) or attitude. The innate nature of aptitude is in contrast to achievement, which represents knowledge or ability that is gained through learning. Aptitude and intelligence quotient are related, and in some ways opposite views of human mental ability. Whereas intelligence quotient sees intelligence as being a single measurable characteristic affecting all mental ability, aptitude refers to one of many different characteristics which can be independent of each other, such as aptitude for military flight, air traffic control, or computer programming. This is more similar to the theory of multiple intelligences.

Concerning a single measurable characteristic affecting all mental ability, analysis of any group of intelligence test scores will nearly always show them to be highly correlated. The U.S. Department of Labour's General Learning Ability, for instance, is determined by combining Verbal, Numerical and Spatial aptitude subtests. In a given person some are low and others high. In the context of an aptitude test the "high" and "low" scores are usually not far apart, because all ability test scores tend to be correlated. Aptitude is better applied intra-individually to determine what tasks a given individual is more skilled at performing. Inter-individual aptitude differences are typically not very significant due to IQ differences. Of course this assumes individuals have not already been pre-screened for aptitude through some other process such as SAT scores, GRE scores, or finishing medical school. Tests that assess learned skills or knowledge are frequently called achievement tests. However, certain tests can assess both types of constructs. An example that leans both ways is the Armed Services Vocational Aptitude Battery (ASVAB), which is given to recruits entering the armed forces of the United States. Another is the SAT, which is designed as a test of aptitude for college in the United States, but has achievement elements. For example, it tests mathematical reasoning, which depends both on innate mathematical ability and education received in mathematics. Aptitude tests are structured systematic ways of evaluating how people perform on tasks or react to different situations. They have standardised methods of administration and scoring with the results quantified and compared with how others have done at the same tests. They are increasingly administered on a computer. Many people have a fear of tests, but these are usually only part of the overall assessment procedure. Quantitative Aptitude and Reasoning are the most important sections in any public sector as well as private sector exams. The quantitative aptitude test measures the numerical ability and accuracy in mathematical calculations.

This well-researched book is designed for the students preparing for the campus recruitment screening test and the undergraduate students for their course in Quantitative Aptitude and Reasoning.

I thank all members of my team who have helped in the preparation of the book. My special thanks go to "Random Publications" who have published the book.

— Dr. Alok Kumar

Contents

Chapter 1

Introduction

An aptitude is a component of a competency to do a certain kind of work at a certain level, which can also be considered "talent". Aptitudes may be physical or mental. Aptitude is not developed knowledge, understanding, learned or acquired abilities (skills) or attitude. The innate nature of aptitude is in contrast to achievement, which represents knowledge or ability that is gained through learning.

Intelligence

Aptitude and intelligence quotient are related, and in some ways opposite views of human mental ability. Whereas intelligence quotient sees intelligence as being a single measurable characteristic affecting all mental ability, aptitude refers to one of many different characteristics which can be independent of each other, such as aptitude for military flight, air traffic control, or computer programming. This is more similar to the theory of multiple intelligences.

Concerning a single measurable characteristic affecting all mental ability, analysis of any group of intelligence test scores will nearly always show them to be highly correlated. The U.S. Department of Labour's General Learning Ability, for instance, is determined by combining Verbal, Numerical and Spatial aptitude subtests. In a given person some are low and others high. In the context of an aptitude test the "high" and "low" scores are usually not far apart, because all ability test scores tend to be correlated. Aptitude is better applied intra-individually to determine what tasks a given individual is more skilled at performing. Inter-individual aptitude differences are typically not very significant due to IQ differences. Of course this assumes individuals have not already been pre-screened for aptitude through some other process such as SAT scores, GRE scores, or finishing medical school.

Combined Aptitude and Knowledge Tests

Tests that assess learned skills or knowledge are frequently called achievement tests. However, certain tests can assess both types of constructs. An example that leans both ways is the Armed Services Vocational Aptitude Battery (ASVAB), which is given to recruits entering the armed forces of the United States. Another is the SAT, which is designed as a test of aptitude for college in the United States, but has achievement elements. For example, it tests mathematical reasoning, which depends both on innate mathematical ability and education received in mathematics.

Aptitude tests can typically be grouped according to the type of cognitive ability they measure:

1. Fluid intelligence: the ability to think and reason abstractly, effectively solve problems and think strategically. It's more commonly known as 'street smarts' or the ability to 'quickly think on your feet'. An example of what employers can learn from your fluid intelligence is your suitability for the role for which you are applying
2. Crystallised intelligence: the ability to learn from past experiences and to apply this learning to work-related situations. Work situations that require crystallised intelligence include producing and analysing written reports, comprehending work instructions, using numbers as a tool to make effective decisions, etc.

Standardized Test

A standardized test is a test that is administered and scored in a consistent, or "standard", manner. Standardized tests are designed in such a way that the questions, conditions for administering, scoring procedures, and interpretations are consistent and are administered and scored in a predetermined, standard manner.

Any test in which the same test is given in the same manner to all test takers is a standardized test. Standardized tests need not be high-stakes tests, time-limited tests, or multiple-choice tests. The opposite of a standardized test is a *non-standardized test.* Non-standardized testing gives significantly different tests to different test takers, or gives the same test under significantly different conditions (e.g., one group is permitted far less time to complete the test than the next group), or evaluates them differently (e.g., the same answer is counted right for one student, but wrong for another student).

Standardized tests are perceived as being more fair than non-standardized tests. The consistency also permits more reliable comparison of outcomes across all test takers.

History

China: The earliest evidence of standardized testing was in China, where the imperial examinations covered the Six Arts which included music, archery and horsemanship, arithmetic, writing, and knowledge of the rituals and ceremonies of both public and private parts. Later, sections on military strategies, civil law, revenue and taxation, agriculture and geography were added to the testing. In this form, the examinations were institutionalized for more than a millennium.

UK: Standardized testing was introduced into Europe in the early 19th century, modelled on the Chinese mandarin examinations, through the advocacy of British colonial administrators, the most "persistent" of which was Britain's consul in Guangzhou, China, Thomas Taylor Meadows. Meadows warned of the collapse of the British Empire if standardized testing was not implemented throughout the empire immediately.

Prior to their adoption, standardized testing was not traditionally a part of Western pedagogy; based on the skeptical and open-ended tradition of debate inherited from Ancient Greece, Western academia favoured non-standardized assessments using essays written by students. It is because of this that the first European implementation of standardized testing did not occur in Europe proper, but in British India. Inspired by the Chinese use of standardized testing, in the early 19th century, British "company managers hired and promoted employees based on competitive examinations in order to prevent corruption and favouritism." This practice of standardized testing was later adopted in the late 19th century by the British mainland. The parliamentary debates that ensued made many references to the "Chinese mandarin system."

It was from Britain, that standardized testing spread, not only throughout the British Commonwealth, but to Europe and then America. Its spread was fueled by the Industrial Revolution. Given the large number of school students during and after the Industrial Revolution, when compulsory education laws increased student populations, open-ended assessment of all students decreased. Moreover, the lack of a standardized process introduces a substantial source of measurement error, as graders might show favouritism or might disagree with each other about the relative merits of different answers.

More recently, it has been shaped in part by the ease and low cost of grading of multiple-choice tests by computer. Grading essays by computer is more difficult, but is also done. In other instances, essays and other open-ended responses are graded according to a pre-determined assessment rubric by trained graders.

United States: The use of standardized testing in the United States is a 20th-century phenomenon with its origins in World War I and the Army Alpha and Beta tests developed by Robert Yerkes and colleagues. Contributing to the growth of standardized tests in the United States in the mid 1800s was immigration. Standardized tests were used in immigration when people first came over to test social roles and find social power and status.

Another example of standardized testing and how it started in the United States was by Everett Lindquist, a professor from Iowa, who made the ACT (American College Testing), which is a very well known standardized test. The ACT includes 4 main sections with multiple choice questions, these sections include English, Math, Reading, and Science, with an optional writing section.

The need for the federal government to make meaningful comparisons across a highly de-centralized (locally controlled) public education system has also contributed to the debate about standardized testing, including the Elementary and Secondary Education Act of 1965 that required standardized testing in public schools. U.S. Public Law 107-110, known as the No Child Left Behind Act of 2001, further ties public school funding to standardized testing.

Standardized testing is a very common way of determining the intelligence of a student in this generation. Ever since 2002, when the No Child Left Behind Act was signed by George Bush ("No Child Left Behind"), students have been given yearly tests which are used to determine the student's intelligence within a short amount of time. Many people disagree with standardized testing. These people argue that standardized testing causes extremely high pressure/stress to do well. The ACT and/or SAT are used in college admissions, so that causes a student with expectations of getting into college to feel pressure to do well on these tests. The difference of a couple numbers of his/her score can mean he/she was accepted, or was not. Another way that some people found that standardized testing is not useful is the fact that many teachers are more focused on teaching to the test than they are teaching the students the material actually needed to know ("Problems with Standardized Testing"). Having a group of students do well on their tests will reflect as if he/she is a good teacher. In order to ensure that the students perform well, the teacher will spend a majority of class time teaching to the test.

Design and Scoring

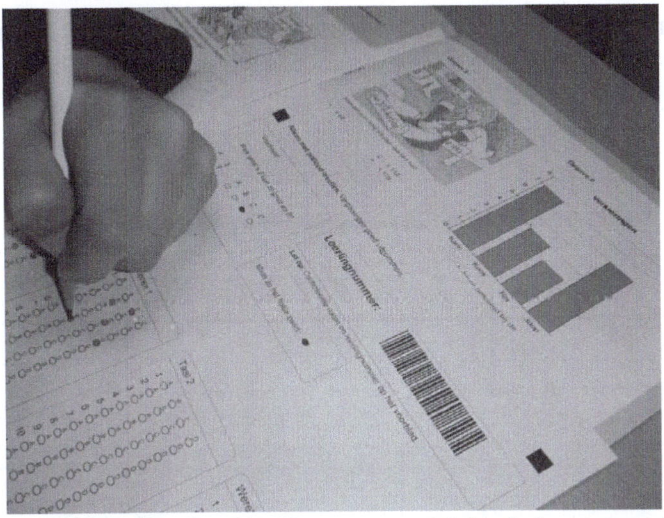

Figure: *Some standardized testing uses multiple-choice tests, which are relatively inexpensive to score, but any form of assessment can be used.*

Standardized testing can be composed of multiple-choice questions, true-false questions, essay questions, authentic assessments, or nearly any other form of assessment. Multiple-choice and true-false items are often chosen because they can be given and scored inexpensively and quickly by scoring special answer sheets by computer or via computer-adaptive testing. Some standardized tests have short-answer or essay writing components that are assigned a score by independent evaluators who use rubrics (rules or guidelines) and benchmark papers (examples of papers for each possible score) to determine the grade to be given to a response. Not all standardized tests involve answering questions; an authentic assessment for athletic skills could take the form of running for a set amount of time or dribbling a ball for a certain distance.

Most assessments, however, are not scored by people; people are used to score items that are not able to be scored easily by computer (i.e., essays). For example, the Graduate Record Exam is a computer-adaptive assessment that requires no scoring by people except for the writing portion.

Scoring Issues

Human scoring is often variable, which is why computer scoring is preferred when feasible. For example, some believe that poorly paid employees will score tests badly. Agreement between scorers can vary between 60 to 85 percent, depending on the test and the scoring session. Sometimes states pay to have two or more scorers read each paper; if their scores do not agree, then the paper is passed to additional scorers.

Open-ended components of tests are often only a small proportion of the test. Most commonly, a major test includes both human-scored and computer-scored sections. These major tests do not measure the student's overall ability in learning.

Score

Sample scoring for the history question: What caused World War II?		
Student answers	Standardized grading	Non-standardized grading
	Grading rubric: Answers must be marked correct if they mention at least one of the following: Germany's invasion of Poland, Japan's invasion of China, or economic issues.	No grading standards. Each teacher grades however he/she wants to, considering factors like the answer, the student's academic potential, and attitude.
Student #1: WWII was caused by Hitler and Germany invading Poland.	*Teacher #1:* This answer mentions one of the required items, so it is correct. *Teacher #2:* This answer is correct.	*Teacher #1:* I feel like this answer is good enough, so I'll mark it correct. *Teacher #2:* This answer is correct, but this good student should be able to do better than that, so I'll only give partial credit.
Student #2: WWII was caused by multiple factors, including the Great Depression and the general economic situation, the rise of nationalism, fascism, and imperialist expansionism; and unresolved resentments related to WWI. The war in Europe began with the German invasion of Poland.	*Teacher #1:* This answer mentions one of the required items, so it is correct. *Teacher #2:* This answer is correct.	*Teacher #1:* I feel like this answer is correct and complete, so I'll give full credit. *Teacher #2:* I feel like this answer is correct, so I'll give full points.
Student #3: WWII was caused by the assassination of Archduke Ferdinand.	*Teacher #1:* This answer does not mention any of the required items. No points. *Teacher #2:* This answer is wrong. No credit.	*Teacher #1:* This answer is wrong. No points. *Teacher #2:* This answer is wrong, but this student tried hard and the sentence is grammatically correct, so I'll give one point for effort.

Standardized testing can be composed of multiple-choice questions, true-false questions, essay questions, authentic assessments, or nearly any other form of assessment. Multiple-choice and true-false items are often chosen because they can be given and scored inexpensively and quickly by scoring special answer sheets by computer or via computer-adaptive testing. Some standardized tests have short-answer or essay writing components that are assigned a score by independent evaluators who use rubrics (rules or guidelines) and benchmark papers (examples of papers for each possible score) to determine the grade to be given to a response. Not all standardized tests involve answering questions; an authentic assessment for athletic skills could take the form of running for a set amount of time or or dribbling a ball for a certain distance.

Most assessments, however, are not scored by people; people are used to score items that are not able to be scored easily by computer (i.e., essays). For example, the Graduate Record Exam is a computer-adaptive assessment that requires no scoring by people except for the writing portion.

Scoring Issues

Human scoring is often variable, which is why computer scoring is preferred when feasible. For example, some believe that poorly paid employees will score tests badly. Agreement between scorers can vary between 60 to 85 percent, depending on the test and the scoring session. Sometimes states pay to have two or more scorers read each paper; if their scores do not agree, then the paper is passed to additional scorers.

Open-ended components of tests are often only a small proportion of the test. Most commonly, a major test includes both human-scored and computer-scored sections. These major tests do not measure the student's overall ability in learning.

Score

There are two types of standardized test score interpretations: a norm-referenced score interpretation or a criterion-referenced score interpretation.

- Norm-referenced score interpretations compare test-takers to a sample of peers. The goal is to rank students as being better or worse than other students. Norm-referenced test score interpretations are associated with traditional education. Students who perform better than others pass the test, and students who perform worse than others fail the test.
- Criterion-referenced score interpretations compare test-takers to a criterion (a formal definition of content), regardless of the scores of other examinees. These may also be described as standards-based assessments, as they are aligned with the standards-based education reform movement. Criterion-referenced score interpretations are concerned solely with whether or not this particular student's answer is correct and complete. Under criterion-referenced systems, it is possible for all students to pass the test, or for all students to fail the test.

Either of these systems can be used in standardized testing. What is important to standardized testing is whether all students are asked equivalent questions, under equivalent circumstances, and graded equally. In a standardized test, if a given answer is correct for one student, it is correct for all students. Graders do not accept an answer as good enough for one student but reject the same answer as inadequate for another student.

Standards

The considerations of validity and reliability typically are viewed as essential elements for determining the quality of any standardized test. However, professional and practitioner associations frequently have placed these concerns within broader contexts when developing standards and making overall judgments about the quality of any standardized test as a whole within a given context.

Evaluation Standards

In the field of evaluation, and in particular educational evaluation, the Joint Committee on Standards for Educational Evaluation has published three sets of standards for evaluations. *The Personnel Evaluation Standards* was published in 1988, *The Program Evaluation Standards* was published in 1994, and *The Student Evaluation Standards* was published in 2003.

Each publication presents and elaborates a set of standards for use in a variety of educational settings. The standards provide guidelines for designing, implementing, assessing and improving the identified form of evaluation. Each of the standards has been placed in one of four fundamental categories to promote educational evaluations that are proper, useful, feasible, and accurate. In these sets of standards, validity and reliability considerations are covered under the accuracy topic. For example, the student accuracy standards help ensure that student evaluations will provide sound, accurate, and credible information about student learning and performance.

Testing Standards

In the field of psychometrics, the Standards for Educational and Psychological Testing place standards about validity and reliability, along with errors of measurement and issues related to the accommodation of individuals with disabilities. The third and final major topic covers standards related to testing applications, credentialing, plus testing in program evaluation and public policy.

Standards for Educational and Psychological Testing

The Standards for Educational and Psychological Testing is a set of testing standards developed jointly by the American Educational

Research Association (AERA), American Psychological Association (APA), and the National Council on Measurement in Education (NCME).

The new edition of *The Standards for Educational and Psychological Testing* was released in July 2014. Five areas received particular attention in the 2014 revision:

1. Examining accountability issues associated with the uses of tests in educational policy
2. Broadening the concept of accessibility of tests for all examinees
3. Representing more comprehensively the role of tests in the workplace
4. Taking into account the expanding role of technology in testing
5. Improving the structure of the book for better communication of the standards

Overview of Organisation and Content

Part I: Test Construction, Evaluation, and Documentation

1. Validity
2. Reliability and Errors of Measurement
3. Test Development and Revision
4. Scales, Norms, and Score Comparability
5. Test Administration, Scoring, and Reporting
6. Supporting Documentation for Tests

Part II: Fairness in Testing

7. Fairness in Testing and Test Use
8. The Rights and Responsibilities of Test Takers
9. Testing Individuals of Diverse Linguistic Backgrounds
10. Testing Individuals with Disabilities

Part III: Testing Applications

11. The Responsibilities of Test Users
12. Psychological Testing and Assessment
13. Educational Testing and Assessment
14. Testing in Employment and Credentialing
15. Testing in Program Evaluation and Public Policy

Related Standards

In 1974, the Joint Committee on Standards for Educational Evaluation was charged with the responsibility of writing a companion volume to the 1974 revision of the *Standards for Educational and Psychological Tests.* This companion volume was to deal with issues and standards for program and curriculum evaluation in education. In 1975, the Joint Committee began work and ultimately decided to establish three separate sets of standards. These standards include The Personnel Evaluation Standards, The Program Evaluation Standards, and The Student Evaluation Standards.

Advantages

One of the main advantages of standardized testing is that the results can be empirically documented; therefore, the test scores can be shown to have a relative degree of validity and reliability, as well as results which are generalizable and replicable. This is often contrasted with grades on a school transcript, which are assigned by individual teachers. It may be difficult to account for differences in educational culture across schools, difficulty of a given teacher's curriculum, differences in teaching style, and techniques and biases that affect grading. This makes standardized tests useful for admissions purposes in higher education, where a school is trying to compare students from across the nation or across the world.

Another advantage is aggregation. A well designed standardized test provides an assessment of an individual's mastery of a domain of knowledge or skill which at some level of aggregation will provide useful information. That is, while individual assessments may not be accurate enough for practical purposes, the mean scores of classes, schools, branches of a company, or other groups may well provide useful information because of the reduction of error accomplished by increasing the sample size.

Standardized tests, which by definition give all test-takers the same test under the same (or reasonably equal) conditions, are also perceived as being more fair than assessments that use different questions or different conditions for students according to their race, socioeconomic status, or other considerations.

Effects

United States: Standardized tests are used in every school around the United States in nearly every grade level. These tests are referred to as high stakes testing and come with many names such as Iowa Tests of Basic Skills, ACT, and SAT; however they all serve the same purpose. All of the testing given in this manner is used to judge the performance of the nations' students and determine their proficiency amongst their peers. Teachers are also measured based on students' results on standardized tests. If a student is found to be less than average it is said to reflect on the teacher and his/her abilities. It is with these perceptions that the United States puts its students in danger. The other problem with SAT, and ACT, is that the tests don't test people who are talented in other domains such as, art, athletics, creative writing and many others.

Testing in schools is used in a wide variety of ways: placing children into learning groups, ranking schools amongst others in the region, state, and nation, and creating a visual for where the United States as a whole is heading. What surprises many is standardized testing may also be a way schools determine merit pay for teachers. Teachers in all grade levels are encouraged to shape their classroom around the upcoming test in hopes that their students outperform others. The effects of this kind of teaching are not beneficial to anyone, except potentially the teacher whose students do well. In the article "Standardized Testing and Its Victims" author Alfie Kohn states, "Schools across the country are cutting back or even eliminating programs in the arts, recess for young children, electives for high schoolers, class meetings (and other activities intended to promote social and moral learning), discussions about current events (since that material will not appear on the test), the use of literature in the early grades (if the tests are focused narrowly on decoding skills), and entire subject areas such as science (if the tests only cover language arts and math)".

Cutting back on real classroom learning is taking its toll on teachers who were genuinely interested in reaching out to the youth and helping them grow. "Many educators are leaving the field because of what is being done to schools in the name of 'accountability' and 'tougher standards'". Teachers are becoming displeased with the field and the ones who genuinely care about student growth are abdicating their roles as educators simply because it has become a twisted version of what it used to be. Prospective educators are now second guessing their choice of careers due to the pressure that will be put upon them to produce the high test scores that matter the most to their potential employers.

With all the stress teachers and administrators are under it would be unreasonable to think it does not rub off on the students as well. Some schools go as far as putting up a visual aid to show where their students fall compared to their classmates. This allows the students to see which of their classmates are proficient, which can be embarrassing for students who fall below the given line. Teachers have many chances to attain their merit pay; a student may only have one chance to pass a test allowing them to move to the next grade level. A single test can determine the outcome of a student's entire educational career, not doing well can be a detriment to their self-esteem. A fourth grader does not need to feel devalued because of a test, they are still developing at an unsteady pace and expecting them all to fall into a neat category of proficiency is simply not acceptable. "Virtually all specialists condemn the practice of giving standardized tests to children younger than 8 or 9 years old".

Students feel the pressure put upon them in a completely different way than an adult would. When asked if students feel the pressure to achieve higher scores on standardized tests educator, Ashley Grossman, states, "I don't think they fully understand it. They feel intimidated and stressed around test time but some of them are like that with any test" (Grossman). Children cannot feel pressured constantly without it having a negative impact on their emotional and potentially physical state. Stress impacts children much the same as it can an adult, sometimes more severe. According to author Josh Ska, "Symptoms of too much stress are usually very evident in children, although they might be mistaken for being rebellious or difficult. A child who frequently blows up over nothing may be having problems at school or at home which are causing chronic stress. Another possible sign of stress is jumpiness or nervousness and poor concentration, which may affect schoolwork. Children who are stressed out may also stop eating or get sick more frequently. The constant adrenaline rush can keep them awake at night and you might notice that your child seems to have insomnia, although she is exhausted. Stomachaches are a common complaint among children suffering from this problem, as are bowel problems and headaches".

Machines scoring tests do not lessen the bias of testing in any way. For the multiple choice problems it is a simple right or wrong; however, computers have been used to score essay portions as well. Criterion is the name of grading software to determine the proficiency of a student's writing abilities. The University of California was considering using this software to determine if students were eligible to skip a writing course, which the instructor was opposed to. In order to prove his point the instructor, Andy Jones, took a letter of recommendation he had written to score it. Author Alain Jehlen notes, " He replaced the student's name with a few words from a Criterion writing prompt, and substituted 'chimpanzee' for every 'the.' Criterion loved the result, calling it 'cogent' and 'well-articulated'" (Jehlen 3). If changing a single word and creating a nonsensical paper was scored so well, then one can only imagine what kinds of writing samples this machine was letting through and calling "wonderful."

Claims have been brought against standardized tests in court due to bias. The legality of a test is based on seven factors: disparate impact (unjustified adverse impact on members of a protected class), validation studies (tests must be validated), state interest, notice and implementation (due process), judicial deference (deferring to a professional educator), remediation and retakes (the amount of remediation offered and the number of retakes), and if the test is homemade. With all these things taken into account there are still several cases where a test was found to be biased and was ruled unreliable by the judicial system.

Debra P. v. Turlington is a case documented where a standardized test was challenged on the basis of racial bias. The SSAT II was claimed to be unconstitutional in the way it was able to deny the students who failed the test high school diplomas. The students in question were provided inadequate notice of the graduation requirements and not given adequate time to prepare themselves for the test. Shelly Mack notes in her research, "The court found that the SSAT II had a clear disproportionate impact on African American students, and noted that Florida intended to discriminate against African American children between 1967-1971 (when the current graduating class was in school under the dual system)" (Mack 2). The state admitted to knowingly discriminating against these children so they would not graduate.

Crump v. Gilmer Independent School District is another case in which graduation was hanging on a single test. Three students had all failed the Texas Assessment of Academic Skills Examination (TAAS). Two of these students had successfully completed all other graduation requirements, while the third student had not. TAAS had only become a graduation requirement in 1991, two years before this case was presented in court, making the argument the students had insufficient time to prepare for the test, as per Debra P. v. Turlington stated that there be at least four to six years of preparation time from its announcement before a new process could be implemented. The two students who had successfully completed all other graduation requirements were granted their diplomas, while the third student was not. It was deemed that the third student's denial was constitutional because there seemed to be no effort on his part from an academic standpoint.

Despite the biases of standardized testing the question remains of if these tests even show actual learning or learning potential of a student. The answer seems to be a resounding "no" from all sources. The number of guesses that are marked correct do not indicate the student has mastered the skill in question; more often than not they had a one in four chance of being correct. Wrong answers are measured correctly, as the student clearly did not know the material, but the correct answers are not indicative of knowledge. A correct answer can point to two other possibilities than mastery of skill; "A correct answer can be achieved using memorization without any profound understanding of the underlying content or conceptual structure of the problem posed" ("Standardized Test" 2) or simply a blind guess resulting in a positive outcome.

However there are positive aspects to standardized tests; specifically for young children. The purpose of standardized tests for young children is to identify developmental delays and to evaluate a young child's development. The standardized tests used for young children are screening tests, diagnostic tests, language tests, and achievement tests. A screening test is used in order to detect an indication of a developmental problem—it identifies if a problem needs to be investigated further. A diagnostic test is done if a child has already been screen tests and indicates further evaluation. Diagnostic tests are designed to assess developmental problems related to learning disabilities. A language test is often administered to students who are considered at-risk. Language tests determine if a student would benefit from a language enrichment program. The achievement test was designed for children in the Head Start program and was introduced by the George W. Bush administration (Wortham, 2008). Overall standardized tests are not solely used to assess young children but is a great way to detect developmental problems in young children.

One proponent of standardized testing is the No Child Left Behind Act (NCLB). This bill supports standards-based education reform, "the belief that setting high standards and establishing measurable goals can improve individual outcomes in education" ("No Child Left Behind Act" 1). NCLB is what set the testing frenzy of the United States in motion. The national government felt it had to step in and take over where state governments had been failing. All the act seemed to do in reality was set up a system of incentives for educators if test results improved. "The system of incentives and penalties sets up a strong motivation for schools, districts, and states to manipulate test results. For example, schools have been shown to employ 'creative reclassification' of drop-outs (to reduce unfavourable statistics)" ("No Child Left Behind Act" 3).

NCLB has encouraged the "teach to the test" method more and more schools have put into place, which leads to students not properly interpreting the test materials despite having been trained for them. Teachers are taught to anticipate what will be on the test and teach the students only that material, leading to students having vague, if any, understanding of any other concepts they may need. "Many teachers who practice 'teaching to the test' actually misinterpret the educational outcomes the tests are designed to measure. On two state tests (New York State and Michigan) and the National Assessment of Educational Progress (NAEP) almost two-thirds of eighth graders missed math word problems that required an application of the Pythagorean theorem to calculate the distance between two points" ("No Child Left Behind Act" 3).

Standardized testing is a detriment to students, affecting them psychologically, emotionally, and intellectually. Their self-esteem is lowered when they do not receive scores they may be aiming for, or when they do not do as well as their classmates. Students are put under undue stress to outperform, simply because teachers are put under stress to make sure their students do well. Important programs are slowly being taken from schools in order to focus on "teaching to the test." Students should be learning the social and moral skills that come with being in particular extracurricular groups or elective classes along with their basic subjects, but with classrooms being test oriented some of the most important real world skills are being taken away from them. This sends up a very real red flag for the future about the kinds of people that will be running the United States. They may be goal oriented, but being people oriented is just as important quality to have.Part of the blame falls to the educators, administrators, and states for not speaking out, but most of the blame lies with the government for increasing standards in a way that is unhealthy.

Public Policy

Standardized testing is used as a public policy strategy to establish stronger accountability measures for public education. While the National Assessment of Education Progress (NAEP) has served as an educational barometer for some thirty years by administering standardized tests on a regular basis to random schools throughout the United States, efforts over the last decade at the state and federal levels have mandated annual standardized test administration for all public schools across the country.

The idea behind the standardized testing policy movement is that testing is the first step to improving schools, teaching practice, and educational methods through data collection. Proponents argue that the data generated by the standardized tests act like a 'report card' for the community, demonstrating how well local schools are performing. Critics of the movement, however, point to various discrepancies that result from current state standardized testing practices, including problems with test validity and reliability and false correlations.

Critics charge that standardized tests became a mandatory curriculum placed into schools without public debate and without any accountability measures of its own. Many feel this ignores basic democratic principles in that control of schools' curricula is removed from local school boards, which are the nominal curricular authority in the U.S. While some maintain that it would be preferable to simply introduce mandatory national curricula, others feel that state mandated standardized testing should stop altogether in order that schools can focus their efforts on instructing their students as they see fit.

Critics also charge that standardized tests encourage "teaching to the test" at the expense of creativity and in-depth coverage of subjects not on the test. Multiple choice tests are criticized for failing to assess skills such as writing. Furthermore, student's success is being tracked to a teacher's relative performance, making teacher advancement contingent upon a teacher's success with a student's academic performance. Ethical and economical questions arise for teachers when faced with clearly underperforming or underskilled students and a standardized test.

Disadvantages and Criticism

The validity, efficacy, and predictive power of standardized tests is disputed. Many contend that overuse and misuse of these tests harms teaching and learning by narrowing the curriculum. According to the group FairTest, when standardized tests are the primary factor in accountability, schools use the tests to narrowly define curriculum and focus instruction. FairTest says that negative consequences of test misuse include narrowing the curriculum, teaching to the test, pushing students out of school, driving teachers out of the profession, and undermining student engagement and school climate. Critics say that "teaching to the test" disfavours higher-order learning. While it is possible to use a standardized test without letting its contents determine curriculum and instruction, frequently, what is not tested is not taught, and how the subject is tested often becomes a model for how to teach the subject.

Uncritical use of standardized test scores to evaluate teacher and school performance is inappropriate, because the students' scores are influenced by three things: what students learn in school, what students learn outside of school, and the students' innate intelligence. The school only has control over one of these three factors. Value-added modelling has been proposed to cope with this criticism by statistically controlling for innate ability and out-of-school contextual factors. In a value-added system of interpreting test scores, analysts estimate an expected score for each student, based on factors such as the student's own previous test scores, primary language, or socioeconomic status. The difference between the student's expected score and actual score is presumed to be due primarily to the teacher's efforts.

Supporters of standardized testing respond that these are not reasons to abandon standardized testing in favour of either non-standardized testing or of no assessment at all, but rather criticisms of poorly designed testing regimes. They argue that testing does and should focus educational resources on the most important aspects of education — imparting a pre-defined set of knowledge and skills — and that other aspects are either less important, or should be added to the testing scheme.

In her book, *Now You See It*, Cathy Davidson criticizes standardized tests. She describes our youth as "assembly line kids on an assembly line model," meaning the use of standardized test as a part of a one-size-fits-all educational model. She also criticizes the narrowness of skills being tested and labelling children without these skills as failures or as students with disabilities. Widespread and organised cheating has been a growing culture in today's reformation of schools.

Education theorist Bill Ayers has commented on the limitations of the standardized test, writing that "Standardized tests can't measure initiative, creativity, imagination, conceptual thinking, curiosity, effort, irony, judgment, commitment, nuance, good will, ethical reflection, or a host of other valuable dispositions and attributes. What they can measure and count are isolated skills, specific facts and function, content knowledge, the least interesting and least significant aspects of learning."

In an April 1995 "meta-analysis" published in the Journal of Educational and Psychological Measurement, Todd Morrison and Melanie Morrison examined two dozen validity studies of the test required to get into just about any Masters or PhD program in America: the Graduate Record Examination (GRE). This study encompassed more than 5,000 test-takers over the past 30 years. The authors found that GRE scores accounted for just 6 percent of the variation in grades in graduate school. The GRE appears to be "virtually useless from a prediction standpoint," wrote the authors. Repeated studies of the Law

School Admissions Test (LSAT) find the same. The SAT's maker, the Educational Testing Service (ETS), now claims the SAT is not an "aptitude" test but rather an assessment of "developed abilities."

Another disadvantage to Standardized Tests is the cost. It has been reported that the United States spends about 1.7 billion dollars annually on these tests.

Scoring Information Loss

A test question might require a student to calculate the area of a triangle. Compare the information provided in these two answers.

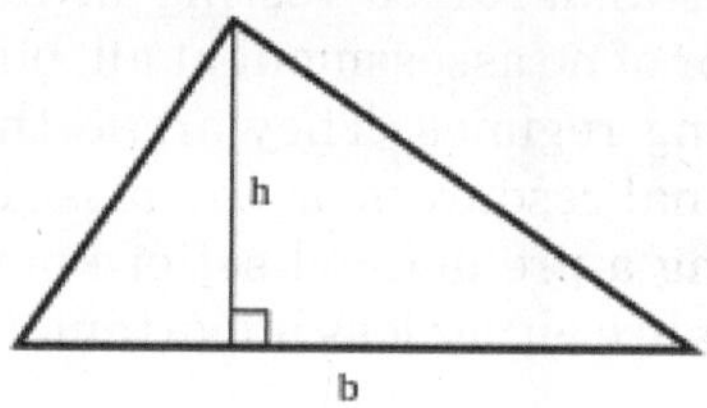

Area = 7.5 cm^2

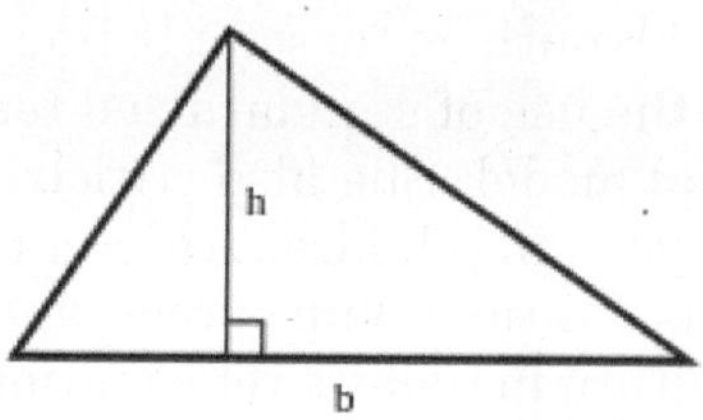

Base = 5 cm; Height = 3 cm

Area = $^1/_2$(Base × Height)

= $^1/_2$(5 cm × 3 cm)

= 7.5 cm^2

Figure: *The first shows scoring information loss. The teacher knows whether the student got the right answer, but does not know how the student arrived at the answer. If the answer is wrong, the teacher does not know whether the student was guessing, made a simple error, or fundamentally misunderstands the subject.*

When tests are scored *right-wrong*, an important assumption has been made about learning. The number of *right* answers or the sum of item scores (where partial credit is given) is assumed to be the appropriate and sufficient measure of current performance status. In addition, a secondary assumption is made that there is no meaningful information in the *wrong* answers.

In the first place, a correct answer can be achieved using *memorization* without any profound understanding of the underlying content or conceptual structure of the problem posed. Second, when more than one step for solution is required, there are often a variety of approaches to answering that will lead to a *correct* result. The fact that the answer is correct does not indicate which of the several possible procedures were used. When the student supplies the answer (or shows the work) this information is readily available from the original documents.

Second, if the *wrong* answers were *blind* guesses, there would be no information to be found among these answers. On the other hand, if *wrong* answers reflect interpretation departures from the expected one, these answers should show an ordered relationship to whatever the overall test is measuring. This departure should be dependent upon the level of psycholinguistic maturity of the student choosing or giving the answer in the vernacular in which the test is written.

In this second case it should be possible to extract this order from the responses to the test items. Such extraction processes, the Rasch model for instance, are standard practice for item development among professionals. However, because the *wrong* answers are discarded during the scoring process, attempts to interpret these answers for the information they might contain is seldom undertaken.

Third, although topic-based subtest scores are sometimes provided, the more common practice is to report the total score or a rescaled version of it. This rescaling is intended to compare these scores to a standard of some sort. This further collapse of the test results systematically removes all the information about which particular items were missed.

Thus, scoring a test *right–wrong* loses 1) how students achieved their *correct* answers, 2) what led them astray towards unacceptable answers and 3) where within the body of the test this departure from expectation occurred.

This commentary suggests that the current scoring procedure conceals the dynamics of the test-taking process and obscures the capabilities of the students being assessed. Current scoring practice oversimplifies these data in the initial scoring step. The result of this procedural error is to obscure of the diagnostic information that could help teachers serve their students better. It further prevents those who are diligently preparing these tests from being able to observe the information that would otherwise have alerted them to the presence of this error.

A solution to this problem, known as Response Spectrum Evaluation (RSE), is currently being developed that appears to be capable of recovering all three of these forms of information loss, while still providing a numerical scale to establish current performance status and to track performance change.

This RSE approach provides an interpretation of the thinking processes behind every answer (both the right and the wrong ones) that tells teachers how they were thinking for every answer they provide. Among other findings, this chapter reports that the recoverable information explains between two and three times more of the test variability than considering only the right answers. This massive loss of information can be explained by the fact that the "wrong" answers are removed from the test information being collected during the scoring process and is no longer available to reveal the procedural error inherent in right-wrong scoring. The procedure bypasses the limitations produced by the linear dependencies inherent in test data.

Testing bias occurs when a test systematically favours one group over another, even though both groups are equal on the trait the test measures. Critics allege that test makers and facilitators tend to represent a middle class, white background. Critics claim that standardized testing match the values, habits, and language of the test makers. However, being that most tests come from a white, middle-class background, it is important to note that the highest scoring groups are not people of that background, but rather tend to come from Asian populations.

Not all tests are well-written, for example, containing multiple-choice questions with ambiguous answers, or poor coverage of the desired curriculum. Some standardized tests include essay questions, and some have criticized the effectiveness of the grading methods. Recently, partial computerized grading of essays has been introduced for some tests, which is even more controversial.

Educational Decisions

Test scores are in some cases used as a sole, mandatory, or primary criterion for admissions or certification. For example, some U.S. states require high school graduation examinations. Adequate scores on these exit exams are required for high school graduation. The General Educational Development test is often used as an alternative to a high school diploma.

Other applications include tracking (deciding whether a student should be enrolled in the "fast" or "slow" version of a course) and awarding scholarships. In the United States, many colleges and universities automatically translate scores on Advanced Placement tests into college credit, satisfaction of graduation requirements, or placement in more advanced courses. Generalized tests such as the SAT or GRE are more often used as one measure among several, when making admissions decisions. Some public institutions have cutoff scores for the SAT, GPA, or class rank, for creating classes of applicants to automatically accept or reject.

Heavy reliance on standardized tests for decision-making is often controversial, for the reasons noted above. Critics often propose emphasizing cumulative or even non-numerical measures, such as classroom grades or brief individual assessments (written in prose) from teachers. Supporters argue that test scores provide a clear-cut, objective standard that minimizes the potential for political influence or favouritism.

The National Academy of Sciences recommends that major educational decisions not be based solely on a test score. The use of minimum cut-scores for entrance or graduation does not imply a single standard, since test scores are nearly always combined with other minimal criteria such as number of credits, prerequisite courses, attendance, etc. Test scores are often perceived as the "sole criteria" simply because they are the most difficult, or the fulfillment of other criteria is automatically assumed. One exception to this rule is the GED, which has allowed many people to have their skills recognised even though they did not meet traditional criteria.

Chapter 2

Ratio, Proportion, Indices and Logarithm

In mathematics, a ratio is a relationship between two numbers of the same kind (*e.g.*, objects, persons, students, spoonfuls, units of whatever identical dimension), expressed as "*a* to *b*" or *a*:*b*, sometimes expressed arithmetically as a dimensionless quotient of the two that explicitly indicates how many times the first number contains the second (not necessarily an integer).

Figure: *The ratio of width to height of standard-definition television.*

In layman's terms a ratio represents, for every amount of one thing, how much there is of another thing. For example, supposing one has 8 oranges and 6 lemons in a bowl of fruit, the ratio of oranges to lemons would be 4:3 (which is equivalent to 8:6) while the ratio of lemons to oranges would be 3:4. Additionally, the ratio of oranges to the total amount of fruit is 4:7 (equivalent to 8:14). The 4:7 ratio can be further converted to a fraction of 4/7 to represent how much of the fruit is oranges.

The ratio of numbers *A* and *B* can be expressed as:

- the ratio of A to B
- A is to B *(often followed by "as ...")*
- A:B
- A fraction (rational number) that is the quotient A divided by B: $\frac{A}{B}$

The numbers *A* and *B* are sometimes called *terms* with *A* being the antecedent and *B* being the consequent.

The proportion expressing the equality of the ratios *A*:*B* and *C*:*D* is written *A*:*B* = *C*:*D* or *A*:*B*::*C*:*D*. This latter form, when spoken or written in the English language, is often expressed as

A is to *B* as *C* is to *D*.

A, *B*, *C* and *D* are called the terms of the proportion. *A* and *D* are called the *extremes*, and *B* and *C* are called the *means*. The equality of three or more proportions is called a continued proportion.

Ratios are sometimes used with three or more terms. The ratio of the dimensions of a "two by four" that is ten inches long is 2:4:10. A good concrete mix is sometimes quoted as 1:2:4 for the ratio of cement to sand to gravel.

For a mixture of 4/1 cement to water, it could be said that the ratio of cement to water is 4:1, that there is 4 times as much cement as water, or that there is a quarter (1/4) as much water as cement..

Older televisions have a 4:3 aspect ratio, which means that the width is 4/3 of the height; modern widescreen TVs have a 16:9 aspect ratio.

History and Etymology

It is impossible to trace the origin of the *concept* of ratio, because the ideas from which it developed would have been familiar to preliterate cultures. For example, the idea of one village being twice as large as another is so basic that it would have been understood in prehistoric society. However, it is possible to trace the origin of the word "ratio" to the Ancient Greek λüγiò (logos). Early translators rendered this into Latin as ratio ("reason"; as in the word "rational"). (A rational number may be expressed as the quotient of two integers.) A more modern interpretation of Euclid's meaning is more akin to computation or reckoning. Medieval writers used the word *proportio* ("proportion") to indicate ratio and *proportionalitas* ("proportionality") for the equality of ratios.

Euclid collected the results appearing in the Elements from earlier sources. The Pythagoreans developed a theory of ratio and proportion as applied to numbers. The Pythagoreans' conception of number included only what would today be called rational numbers, casting doubt on the validity of the theory in geometry where, as the Pythagoreans also discovered, incommensurable ratios (corresponding to irrational numbers) exist. The discovery of a theory of ratios that does not assume commensurability is probably due to Eudoxus. The exposition of the theory of proportions that appears in Book VII of The Elements reflects the earlier theory of ratios of commensurables.

The existence of multiple theories seems unnecessarily complex to modern sensibility since ratios are, to a large extent, identified with quotients. This is a comparatively recent development however, as can be seen from the fact that modern geometry textbooks still use distinct terminology and notation for ratios and quotients. The reasons for this are twofold. First, there was the previously mentioned reluctance to accept irrational numbers as true numbers. Second, the lack of a widely used symbolism to replace the already established terminology of ratios delayed the full acceptance of fractions as alternative until the 16th century.

Euclid's Definitions

Book V of Euclid's Elements has 18 definitions, all of which relate to ratios. In addition, Euclid uses ideas that were in such common usage that he did not include definitions for them. The first two definitions say that a *part* of a quantity is another quantity that "measures" it and conversely, a *multiple* of a quantity is another quantity that it measures. In modern terminology, this means that a multiple of a quantity is that quantity multiplied by an integer greater than one—and a part of a quantity (meaning aliquot part) is a part that, when multiplied by an integer greater than one, gives the quantity.

Euclid does not define the term "measure" as used here, However, one may infer that if a quantity is taken as a unit of measurement, and a second quantity is given as an integral number of these units, then the first quantity *measures* the second. Note that these definitions are repeated, nearly word for word, as definitions 3 and 5 in book VII.

Definition 3 describes what a ratio is in a general way. It is not rigorous in a mathematical sense and some have ascribed it to Euclid's editors rather than Euclid himself. Euclid defines a ratio as between two quantities *of the same type*, so by this definition the ratios of two lengths or of two areas are defined, but not the ratio of a length and an area. Definition 4 makes this more rigorous. It states that a ratio of two quantities exists when there is a multiple of each that exceeds the other. In modern notation, a ratio exists between quantities p and q if there exist integers m and n so that $mp>q$ and $nq>m$. This condition is known as the Archimedean property.

Definition 5 is the most complex and difficult. It defines what it means for two ratios to be equal. Today, this can be done by simply stating that ratios are equal when the quotients of the terms are equal, but Euclid did not accept the existence of the quotients of incommensurables, so such a definition would have been meaningless to him. Thus, a more subtle definition is needed where quantities involved are not measured directly to one another. Though it may not be possible to assign a rational value to a ratio, it is possible to compare a ratio with a rational number. Specifically, given two quantities, p and q, and a rational number m/n we can say that the ratio of p to q is less than, equal to, or greater than m/n when np is less than, equal to, or greater than mq respectively. Euclid's definition of equality can be stated as that two ratios are equal when they behave identically with respect to being less than, equal to, or greater than any rational number. In modern notation this says that given quantities p, q, r and s, then $p:q::r:s$ if for any positive integers m and n, $np<mq$, $np=mq$, $np>mq$ according as $nr<ms$, $nr=ms$, $nr>ms$ respectively. There is a remarkable similarity between this definition and the theory of Dedekind cuts used in the modern definition of irrational numbers.

Definition 6 says that quantities that have the same ratio are *proportional* or *in proportion.* Euclid uses the Greek νάλϋγïν (analogon), this has the same root as ëüγïò and is related to the English word "analog".

Definition 7 defines what it means for one ratio to be less than or greater than another and is based on the ideas present in definition 5. In modern notation it says that given quantities p, q, r and s, then $p:q>r:s$ if there are positive integers m and n so that $np>mq$ and nrd”ms.

As with definition 3, definition 8 is regarded by some as being a later insertion by Euclid's editors. It defines three terms p, q and r to be in proportion when $p:q::q:r$. This is extended to 4 terms p, q, r and s as $p:q::q:r::r:s$, and so on. Sequences that have the property that the ratios of consecutive terms are equal are called Geometric progressions. Definitions 9 and 10 apply this, saying that if p, q and r are in proportion then $p:r$ is the *duplicate ratio* of $p:q$ and if p, q, r and s are in proportion then $p:s$ is the *triplicate ratio* of $p:q$. If p, q and r are in proportion then q is called a *mean proportional* to (or the geometric mean of) p and r.

Similarly, if p, q, r and s are in proportion then q and r are called two mean proportionals to p and s.

Fraction

If there are 2 oranges and 3 apples, the ratio of oranges to apples is 2:3, and the ratio of oranges to the total number of pieces of fruit is 2:5. These ratios can also be expressed in fraction form: there are 2/3 as many oranges as apples, and 2/5 of the pieces of fruit are oranges. If orange juice concentrate is to be diluted with water in the ratio 1:4, then one part of concentrate is mixed with four parts of water, giving five parts total; the amount of orange juice concentrate is 1/4 the amount of water, while the amount of orange juice concentrate is 1/5 of the total liquid. In both ratios and fractions, it is important to be clear what is being compared to what, and beginners often make mistakes for this reason.

Number of Terms

In general, when comparing the quantities of a two-quantity ratio, this can be expressed as a fraction derived from the ratio. For example, in a ratio of 2:3, the amount/size/volume/number of the first quantity is $\frac{2}{3}$ that of the second quantity. This pattern also works with ratios with more than two terms. However, a ratio with more than two terms cannot be completely converted into a single fraction; a single fraction represents only one part of the ratio since a fraction can only compare two numbers. If the ratio deals with objects or amounts of objects, this is often expressed as "for every two parts of the first quantity there are three parts of the second quantity".

Percentage Ratio

If we multiply all quantities involved in a ratio by the same number, the ratio remains valid. For example, a ratio of 3:2 is the same as 12:8. It is usual either to reduce terms to the lowest common denominator, or to express them in parts per hundred (percent).

If a mixture contains substances A, B, C & D in the ratio 5:9:4:2 then there are 5 parts of A for every 9 parts of B, 4 parts of C and 2 parts of D. As 5+9+4+2=20, the total mixture contains 5/20 of A (5 parts out of 20), 9/20 of B, 4/20 of C, and 2/20 of D. If we divide all numbers by the total and multiply by 100, this is converted to percentages: 25% A, 45% B, 20% C, and 10% D (equivalent to writing the ratio as 25:45:20:10).

Proportions

If the two or more ratio quantities encompass all of the quantities in a particular situation, for example two apples and three oranges in a fruit basket containing no other types of fruit, it could be said that "the whole" contains five parts, made up of two parts apples and three parts oranges. In this case, $\frac{2}{5}$, or 40% of the whole are apples and $\frac{3}{5}$, or 60% of the whole are oranges. This comparison of a specific quantity to "the whole" is sometimes called a proportion. Proportions are sometimes expressed as percentages as demonstrated above.

Reduction

Note that ratios can be reduced (as fractions are) by dividing each quantity by the common factors of all the quantities. This is often called "cancelling." As for fractions, the simplest form is considered that in which the numbers in the ratio are the smallest possible integers.

Thus, the ratio 40:60 may be considered equivalent in meaning to the ratio 2:3 within contexts concerned only with relative quantities.

Mathematically, we write: "40:60" = "2:3" (dividing both quantities by 20).

Grammatically, we would say, "40 to 60 equals 2 to 3."

An alternative representation is: "40:60::2:3"

Grammatically, we would say, "40 is to 60 as 2 is to 3."

A ratio that has integers for both quantities and that cannot be reduced any further (using integers) is said to be in simplest form or lowest terms.

Sometimes it is useful to write a ratio in the form 1:*n* or *n*:1 to enable comparisons of different ratios.

For example, the ratio 4:5 can be written as 1:1.25 (dividing both sides by 4)

Alternatively, 4:5 can be written as 0.8:1 (dividing both sides by 5)

Where the context makes the meaning clear, a ratio in this form is sometimes written without the 1 and the colon, though, mathematically, this makes it a factor or multiplier.

Dilution Ratio

Ratios are often used for simple dilutions applied in chemistry and biology. A simple dilution is one in which a unit volume of a liquid material of interest is combined with an appropriate volume of a solvent liquid to achieve the desired concentration. The dilution factor is the total number of unit volumes in which your material is dissolved. The diluted material must then be thoroughly mixed to achieve the true dilution. For example, a 1:5 dilution (verbalize as "1 to 5" dilution) entails combining 1 unit volume of solute (the material to be diluted) + 4 unit volumes (approximately) of the solvent to give 5 units of the total volume. (Some solutions and mixtures take up slightly less volume than their components.)

The dilution factor is frequently expressed using exponents: 1:5 would be 5e–1 (5^{-1} i.e. one-fifth:one); 1:100 would be 10e–2 (10^{-2} i.e. one hundredth:one), and so on.

There is often confusion between dilution ratio (1:n meaning 1 part solute to n parts solvent) and dilution factor (1:n+1) where the second number (n+1) represents the total volume of solute + solvent. In scientific and serial dilutions, the given ratio (or factor) often means the ratio to the final volume, not to just the solvent. The factors then can easily be multiplied to give an overall dilution factor.

In other areas of science such as pharmacy, and in non-scientific usage, a dilution is normally given as a plain ratio of solvent to solute.

Proportion

In mathematics, two variables are proportional if a change in one is always accompanied by a change in the other, and if the changes are always related by use of a constant multiplier. The constant is called the coefficient of proportionality or proportionality constant.

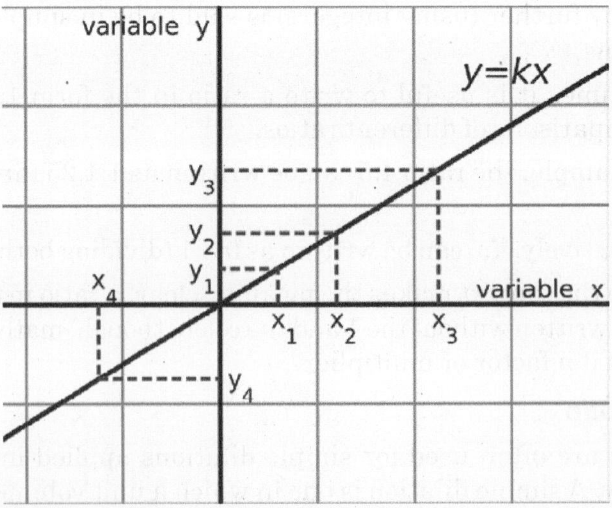

Figure: *Variable y is directly proportional to the variable x.*

- If one variable is always the product of the other and a constant, the two are said to be *directly proportional.* x and y are directly proportional if the ratio $\frac{y}{x}$ is constant.

If the product of the two variables is always equal to a constant, the two are said to be *inversely proportional.* x and y are inversely proportional if the product xy is constant.

To express the statement, "y is proportional to x," we write as an equation y = cx, for some real constant c. Symbolically, we write y "" x.

To express the statement, "y is inversely proportional to x," we write as an equation y = c/x. We can equivalently write, "y is proportional to 1/x", which y = c/x would represent.

If a linear function transforms 0, a and b into 0, c and d, and if the product $a\ b\ c\ d$ is not zero, we say a and b are proportional to c and d. An equality of two ratios such as $\frac{a}{c} = \frac{b}{d}$, where no term is zero, is called a proportion.

Geometric Illustration

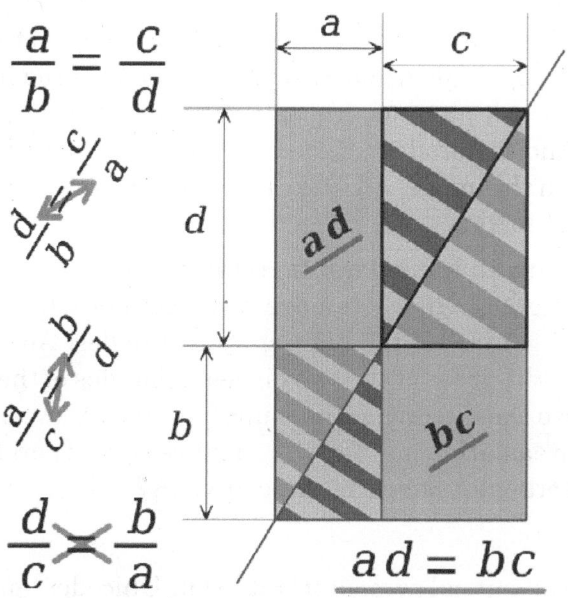

Figure: *The two rectangles with stripes are similar, the ratios of their dimensions are horizontally written within the image. The duplication scale of a striped triangle is obliquely written, in a proportion obtained by inverting two terms of another proportion horizontally written.*

When the duplication of a given rectangle preserves its shape, the ratio of the large dimension to the small dimension is a constant number in all the copies, and in the original rectangle. The largest rectangle of the drawing is similar to one or the other rectangle with stripes. From their width to their height, the coefficient is $\frac{d}{c} = \frac{b}{a} = \frac{d+b}{c+a}$. A ratio of their dimensions horizontally written within the image, at the top or the bottom, determines the common shape of the three similar rectangles.

The common diagonal of the similar rectangles divides each rectangle into two superposable triangles, with two different kinds of stripes. The four striped triangles and the two striped rectangles have a common vertex: the center of an homothetic transformation with a negative ratio $-k$ or $\frac{-1}{k}$, that transforms one triangle and its stripes into another triangle with the same stripes, enlarged or reduced. The duplication scale of a striped triangle is the proportionality constant between the corresponding sides lengths of the triangles, equal to a positive ratio obliquely written within the image:

$$\frac{c}{a} = k \text{ or } \frac{a}{c} = \frac{1}{k}.$$

In the proportion $\frac{a}{b} = \frac{c}{d}$, the terms a and d are called the extremes, while b and c are the means, because a and d are the extreme terms of the list (a, b, c, d), while b and c are in the middle of the list. From any proportion, we get another proportion by inverting the extremes or the means. And the product of the extremes equals the product of the means. Within the image, a double arrow indicates two inverted terms of the first proportion.

Consider dividing the largest rectangle in two triangles, cutting along the diagonal. If we remove two triangles from either half rectangle, we get one of the plain gray rectangles. Above and below this diagonal, the areas of the two biggest triangles of the drawing are equal, because these triangles are superposable. Above and below the subtracted areas are equal for the same reason. Therefore, the two plain gray rectangles have the same area: $a\ d = b\ c$.

Symbols

The mathematical symbol (U+221D in Unicode, \propto in TeX) is used to indicate that two values are proportional. For example, A "" B means the variable A is directly proportional to the variable B.

Other symbols include:

- :: - U+2237 "PROPORTION"
- :: - U+223A "GEOMETRIC PROPORTION"

Direct Proportionality

Given two variables x and y, *y is directly proportional to x* (*x and y vary directly,* or *x an y are in direct variation*) if there is a non-zero constant k such that

$$y = kx.$$

The relation is often denoted, using the :: symbol, as

$$y \propto x$$

and the constant ratio

$$k = \frac{y}{x}$$

is called the proportionality constant, constant of variation or constant of proportionality.

Examples

- If an object travels at a constant speed, then the distance travelled is directly proportional to the time spent travelling, with the speed being the constant of proportionality.
- The circumference of a circle is directly proportional to its diameter, with the constant of proportionality equal to π.
- On a map drawn to scale, the distance between any two points on the map is directly proportional to the distance between the two locations that the points represent, with the constant of proportionality being the scale of the map.
- The force acting on a certain object due to gravity is directly proportional to the object's mass; the constant of proportionality between the mass and the force is known as gravitational acceleration.

Properties

Since

$$y = kx$$

is equivalent to

$$x = \left(\frac{1}{k}\right)y,$$

it follows that if y is directly proportional to x, with (nonzero) proportionality constant k, then x is also directly proportional to y with proportionality constant $1/k$.

If y is directly proportional to x, then the graph of y as a function of x will be a straight line passing through the origin with the slope of the line equal to the constant of proportionality: it corresponds to linear growth.

Inverse Proportionality

The concept of *inverse proportionality* can be contrasted against *direct proportionality*. Consider two variables said to be "inversely proportional" to each other. If all other variables are held constant, the magnitude or absolute value of one inversely proportional variable will decrease if the other variable increases, while their product (the constant of proportionality k) is always the same.

Formally, two variables are inversely proportional (also called varying inversely, in inverse variation, in inverse proportion, in reciprocal proportion) if one of the variables is directly proportional with the multiplicative inverse (reciprocal) of the other, or equivalently if their product is a constant. It follows that the variable y is inversely proportional to the variable x if there exists a non-zero constant k such that

$$y = \frac{k}{x}$$

The constant can be found by multiplying the original x variable and the original y variable.

As an example, the time taken for a journey is inversely proportional to the speed of travel; the time needed to dig a hole is (approximately) inversely proportional to the number of people digging.

The graph of two variables varying inversely on the Cartesian coordinate plane is a hyperbola. The product of the X and Y values of each point on the curve will equal the constant of proportionality (k). Since neither x nor y can equal zero (if k is non-zero), the graph will never cross either axis.

Hyperbolic Coordinates

The concepts of *direct* and *inverse* proportion lead to the location of points in the Cartesian plane by hyperbolic coordinates; the two coordinates correspond to the constant of direct proportionality that locates a point on a ray and the constant of inverse proportionality that locates a point on a hyperbola.

Exponential and Logarithmic Proportionality

A variable y is exponentially proportional to a variable x, if y is directly proportional to the exponential function of x, that is if there exist non-zero constants k and a

$$y = ka^x.$$

Likewise, a variable y is logarithmically proportional to a variable x, if y is directly proportional to the logarithm of x, that is if there exist non-zero constants k and a

$$y = k\log_a(x).$$

Indices

An index is an indirect shortcut derived from and pointing into, a greater volume of values, data, information or knowledge.

In statistics and research design, an index is a composite statistic - a measure of changes in a representative group of individual data points, or in other words, a compound measure that aggregates multiple indicators. Indexes summarize and rank specific observations.

Much data in the field of social sciences is represented in various indices such as Gender Gap Index, Human Development Index or the Dow Jones Industrial Average.

Item in indexes are usually weighted equally, unless there are some reasons against it (for example, if two items reflect essentially the same aspect of a variable, they could have a weight of 0.5 each).

Constructing the items involves four steps. First, items should be selected based on their face validity, unideminsionality, the degree of specificity in which a dimension is to be measured, and their amount of variance. Items should be empirically related to one another, which leads to the second step of examining their multivariate relationships. Third, indexes scores are designed, which involves determining their score ranges and weights for the items. Finally, indexes should be validated, which involves testing whether they can predict indicators related to the measured variable not used in their construction.

Database Index

A database index is a data structure that improves the speed of data retrieval operations on a database table at the cost of additional writes and storage space to maintain the index data structure. Indexes are used to quickly locate data without having to search every row in a database table every time a database table is accessed. Indexes can be created using one or more columns of a database table, providing the basis for both rapid random lookups and efficient access of ordered records.

An index is a copy of select columns of data from a table that can be searched very efficiently that also includes a low level disk block address or direct link to the complete row of data it was copied from. Some databases extend the power of indexing by letting developers create indices on functions or expressions. For example, an index could be created on upper(last_name), which would only store the upper case versions of the last_name field in the index. Another option sometimes supported is the use of partial indices, where index entries are created only for those records that satisfy some conditional expression. A further aspect of flexibility is to permit indexing on user-defined functions, as well as expressions formed from an assortment of built-in functions.

Usage

Support for Fast Lookup: Most database software includes indexing technology that enables sub-linear time lookup to improve performance, as linear search is inefficient for large databases.

Suppose a database contains N data items and one must be retrieved based on the value of one of the fields. A simple implementation retrieves and examines each item according to the test. If there is only one matching item, this can stop when it finds that single item, but if there are multiple matches, it must test everything. This means that the number of operations in the worst case is O(N) or linear time. Since databases commonly contain millions of objects, and since lookup is a common operation, it is often desirable to improve performance.

An index is any data structure that improves the performance of lookup. There are many different data structures used for this purpose, and in fact a substantial proportion of the field of computer science is devoted to the design and analysis of index data structures. There are complex design trade-offs involving lookup performance, index size, and index update performance. Many index designs exhibit logarithmic (O(log(N))) lookup performance and in some applications it is possible to achieve flat (O(1)) performance.

Policing the Database Constraints

Indexes are used to police database constraints, such as UNIQUE, EXCLUSION, PRIMARY KEY and FOREIGN KEY. An index may be declared as UNIQUE, which creates an implicit constraint on the underlying table. Database systems usually implicitly create an index on a set of columns declared PRIMARY KEY, and some are capable of using an already existing index to police this constraint. Many database systems require that both referencing and referenced sets of columns in a FOREIGN KEY constraint are indexed, thus improving performance of inserts, updates and deletes to the tables participating in the constraint.

Some database systems support an EXCLUSION constraint that ensures that, for a newly inserted or updated record, a certain predicate holds for no other record. This can be used to implement a UNIQUE constraint (with equality predicate) or more complex constraints, like ensuring that no overlapping time ranges or no intersecting geometry objects would be stored in the table. An index supporting fast searching for records satisfying the predicate is required to police such a constraint.

Index Architecture/Indexing Methods

Non-Clustered: The data is present in arbitrary order, but the logical ordering is specified by the index. The data rows may be spread throughout the table regardless of the value of the indexed column or expression. The non-clustered index tree contains the index keys in sorted order, with the leaf level of the index containing the pointer to the record (page and the row number in the data page in page-organised engines; row offset in file-organised engines).

In a non-clustered index

- The physical order of the rows is not the same as the index order.
- The indexed columns are typically non-primary key columns used in JOIN, WHERE, and ORDER BY clauses.

There can be more than one non-clustered index on a database table.

Clustered

Clustering alters the data block into a certain distinct order to match the index, resulting in the row data being stored in order. Therefore, only one clustered index can be created on a given database table. Clustered indices can greatly increase overall speed of retrieval, but usually only where the data is accessed sequentially in the same or reverse order of the clustered index, or when a range of items is selected.

Since the physical records are in this sort order on disk, the next row item in the sequence is immediately before or after the last one, and so fewer data block reads are required. The primary feature of a clustered index is therefore the ordering of the physical data rows in accordance with the index blocks that point to them. Some databases separate the data and index blocks into separate files, others put two completely different data blocks within the same physical file(s).

Cluster

When multiple databases and multiple tables are joined, it's referred to as a cluster (not to be confused with clustered index described above). The records for the tables sharing the value of a cluster key shall be stored together in the same or nearby data blocks. This may improve the joins of these tables on the cluster key, since the matching records are stored together and less I/O is required to locate them. The cluster configuration defines the data layout in the tables that are parts of the cluster. A cluster can be keyed with a B-Tree index or a hash table. The data block where the table record is stored is defined by the value of the cluster key.

Column Order

The order that the index definition defines the columns in is important. It is possible to retrieve a set of row identifiers using only the first indexed column. However, it is not possible or efficient (on most databases) to retrieve the set of row identifiers using only the second or greater indexed column.

For example, imagine a phone book that is organised by city first, then by last name, and then by first name. If you are given the city, you can easily extract the list of all phone numbers for that city. However, in this phone book it would be very tedious to find all the phone numbers for a given last name. You would have to look within each city's section for the entries with that last name. Some databases can do this, others just won't use the index.

In the phone book example with a composite index created on the columns (*city, last_name, first_name*), if we search by giving exact values for all the three fields, search time is minimal—but if we provide the values for *city* and *first_name* only, the search uses only the *city* field to retrieve all matched records. Then a sequential lookup checks the matching with *first_name*. So, to improve the performance, one must ensure that the index is created on the order of search columns.

Applications and Limitations

Indexes are useful for many applications but come with some limitations. Consider the following SQL statement: SELECT first_name FROM people WHERE last_name = 'Smith';. To process this statement without an index the database software must look at the last_name column on every row in the table (this is known as a full table scan). With an index the database simply follows the B-tree data structure until the Smith entry has been found; this is much less computationally expensive than a full table scan.

Consider this SQL statement: SELECT email_address FROM customers WHERE email_address LIKE '%@yahoo.com';. This query would yield an email address for every customer whose email address ends with "@yahoo.com", but even if the email_address column has been indexed the database must perform a full index scan. This is because the index is built with the assumption that words go from left to right. With a wildcard at the beginning of the search-term, the database software is unable to use the underlying B-tree data structure (in other words, the WHERE-clause is *not* sargable). This problem can be solved through the addition of another index created on reverse(email_address) and a SQL query like this: SELECT email_address FROM customers WHERE reverse(email_address) LIKE reverse('%@yahoo.com');. This puts the wild-card at the right-most part of the query (now moc.oohay@%), which the index on reverse(email_address) can satisfy.

When the wildcard characters are used on both sides of the search word as *%yahoo.com%*, the index available on this field is not used. Rather only a sequential search is performed, which takes O(N) time. So, index must be available on the columns the lookup is performed on.

Types of Indexes

Bitmap Index: A bitmap index is a special kind of index that stores the bulk of its data as bit arrays (bitmaps) and answers most queries by performing bitwise logical operations on these bitmaps. The most commonly used indexes, such as B+trees, are most efficient if the values they index do not repeat or repeat a smaller number of times. In contrast, the bitmap index is designed for cases where the values of a variable repeat very frequently. For example, the gender field in a customer database usually contains at most three distinct values: male, female or other. For such variables, the bitmap index can have a significant performance advantage over the commonly used trees.

Dense Index

A dense index in databases is a file with pairs of keys and pointers for every record in the data file. Every key in this file is associated with a particular pointer to *a record* in the sorted data file. In clustered indices with duplicate keys, the dense index points *to the first record* with that key.

Sparse Index

A sparse index in databases is a file with pairs of keys and pointers for every block in the data file. Every key in this file is associated with a particular pointer *to the block* in the sorted data file. In clustered indices with duplicate keys, the sparse index points *to the lowest search key* in each block.

Reverse Index

A reverse key index reverses the key value before entering it in the index. E.g., the value 24538 becomes 83542 in the index. Reversing the key value is particularly useful for indexing data such as sequence numbers, where new key values monotonically increase.

Index Implementations

Indices can be implemented using a variety of data structures. Popular indices include balanced trees, B+ trees and hashes.

In Microsoft SQL Server, the leaf node of the clustered index corresponds to the actual data, not simply a pointer to data that resides elsewhere, as is the case with a non-clustered index. Each relation can have a single clustered index and many unclustered indices.

Index Concurrency Control

An index is typically being accessed concurrently by several transactions and processes, and thus needs concurrency control. While in principle indexes can utilize the common database concurrency control methods, specialised concurrency control methods for indexes exist, which are applied in conjunction with the common methods for a substantial performance gain.

Covering Index

In most cases, an index is used to quickly locate the data record(s) from which the required data is read. In other words, the index is only used to locate data records in the table and not to return data.

A covering index is a special case where the index itself contains the required data field(s) and can return the data.

***Table:** Consider the following table (other fields omitted)*

ID	Name	Other Fields
12	Plug	...
13	Lamp	...
14	Fuse	...

To find the Name for ID 13, an index on (ID) is useful, but the record must still be read to get the Name. However, an index on (ID, Name) contains the required data field and eliminates the need to look up the record. A covering index can dramatically speed up data retrieval but may itself be large due to the additional keys, which slow down data insertion & update. To reduce such index size, some systems allow including non-key fields in the index. Non-key fields are not themselves part of the index ordering but only included at the leaf level, allowing for a covering index with less overall index size.

Standardization

No standard defines how to create indexes, because the ISO SQL Standard does not cover physical aspects. Indexes are one of the physical parts of database conception among others like storage (tablespace or filegroups). RDBMS vendors all give a CREATE INDEX syntax with some specific options that depend on their software's capabilities.

Scale

In the social sciences, scaling is the process of measuring or ordering entities with respect to quantitative attributes or traits. For example, a scaling technique might involve estimating individuals' levels of extraversion, or the perceived quality of products. Certain methods of scaling permit estimation of magnitudes on a continuum, while other methods provide only for relative ordering of the entities.

The level of measurement is the type of data that is measured.

The word scale is sometimes (including in academic literature) used to refer to another composite measure, that of an index. Those concepts are however different.

Comparative and Non Comparative Scaling

With comparative scaling, the items are directly compared with each other (example: Do you prefer Pepsi or Coke?). In noncomparative scaling each item is scaled independently of the others (example: How do you feel about Coke?).

Composite Measures

Composite measures of variables are created by combining two or more separate empirical indicators into a single measure. Composite measures measure complex concepts more adequately than single indicators, extend the range of scores available and are more efficient at handling multiple items.

In addition to scales, there are two other types of composite measures. Indexes are similar to scales except multiple indicators of a variable are combined into a single measure. The index of consumer confidence, for example, is a combination of several measures of consumer attitudes. A typology is similar to an index except the variable is measured at the nominal level.

Indexes are constructed by accumulating scores assigned to individual attributes, while scales are constructed through the assignment of scores to patterns of attributes.

While indexes and scales provide measures of a single dimension, typologies are often employed to examine the intersection of two or more dimensions. Typologies are very useful analytical tools and can be easily used as independent variables, although since they are not unidimensional it is difficult to use them as a dependent variable.

Data Types

The type of information collected can influence scale construction. Different types of information are measured in different ways.

1. Some data are measured at the nominal level. That is, any numbers used are mere labels; they express no mathematical properties. Examples are SKU inventory codes and UPC bar codes.
2. Some data are measured at the ordinal level. Numbers indicate the relative position of items, but not the magnitude of difference. An example is a preference ranking.
3. Some data are measured at the interval level. Numbers indicate the magnitude of difference between items, but there is no absolute zero point. Examples are attitude scales and opinion scales.
4. Some data are measured at the ratio level. Numbers indicate magnitude of difference and there is a fixed zero point. Ratios can be calculated. Examples include: age, income, price, costs, sales revenue, sales volume, and market share.

Scale Construction Decisions

- What level of data is involved (nominal, ordinal, interval, or ratio)?
- What will the results be used for?
- Should you use a scale, index, or typology?
- What types of statistical analysis would be useful?
- Should you use a comparative scale or a noncomparative scale?
- How many scale divisions or categories should be used (1 to 10; 1 to 7; –3 to +3)?

- Should there be an odd or even number of divisions? (Odd gives neutral center value; even forces respondents to take a non-neutral position.)
- What should the nature and descriptiveness of the scale labels be?
- What should the physical form or layout of the scale be? (graphic, simple linear, vertical, horizontal)
- Should a response be forced or be left optional?

Comparative Scaling Techniques

- Pairwise comparison scale – a respondent is presented with two items at a time and asked to select one (example : Do you prefer Pepsi or Coke?). This is an ordinal level technique when a measurement model is not applied. Krus and Kennedy (1977) elaborated the paired comparison scaling within their domain-referenced model. The Bradley–Terry–Luce (BTL) model (Bradley and Terry, 1952; Luce, 1959) can be applied in order to derive measurements provided the data derived from paired comparisons possess an appropriate structure. Thurstone's Law of comparative judgment can also be applied in such contexts.
- Rasch model scaling – respondents interact with items and comparisons are inferred between items from the responses to obtain scale values. Respondents are subsequently also scaled based on their responses to items given the item scale values. The Rasch model has a close relation to the BTL model.
- Rank-ordering – a respondent is presented with several items simultaneously and asked to rank them (example : Rate the following advertisements from 1 to 10.). This is an ordinal level technique.
- Bogardus social distance scale – measures the degree to which a person is willing to associate with a class or type of people. It asks how willing the respondent is to make various associations. The results are reduced to a single score on a scale. There are also non-comparative versions of this scale.
- Q-Sort – Up to 140 items are sorted into groups based on rank-order procedure.
- Guttman scale – This is a procedure to determine whether a set of items can be rank-ordered on a unidimensional scale. It utilizes the intensity structure among several indicators of a given variable. Statements are listed in order of importance.

The rating is scaled by summing all responses until the first negative response in the list. The Guttman scale is related to Rasch measurement; specifically, Rasch models bring the Guttman approach within a probabilistic framework.

- Constant sum scale – a respondent is given a constant sum of money, script, credits, or points and asked to allocate these to various items (example : If you had 100 Yen to spend on food products, how much would you spend on product A, on product B, on product C, etc.). This is an ordinal level technique.
- Magnitude estimation scale – In a psychophysics procedure invented by S. S. Stevens people simply assign numbers to the dimension of judgment. The geometric mean of those numbers usually produces a power law with a characteristic exponent. In cross-modality matching instead of assigning numbers, people manipulate another dimension, such as loudness or brightness to match the items. Typically the exponent of the psychometric function can be predicted from the magnitude estimation exponents of each dimension.

Non-Comparative Scaling Techniques

- Continuous rating scale (also called the graphic rating scale) – respondents rate items by placing a mark on a line. The line is usually labelled at each end. There are sometimes a series of numbers, called scale points, (say, from zero to 100) under the line. Scoring and codification is difficult.
- Likert scale – Respondents are asked to indicate the amount of agreement or disagreement (from strongly agree to strongly disagree) on a five- to nine-point scale. The same format is used for multiple questions. This categorical scaling procedure can easily be extended to a magnitude estimation procedure that uses the full scale of numbers rather than verbal categories.
- Phrase completion scales – Respondents are asked to complete a phrase on an 11-point response scale in which 0 represents the absence of the theoretical construct and 10 represents the theorized maximum amount of the construct being measured. The same basic format is used for multiple questions.
- Semantic differential scale – Respondents are asked to rate on a 7 point scale an item on various attributes. Each attribute requires a scale with bipolar terminal labels.
- Stapel scale – This is a unipolar ten-point rating scale. It ranges from +5 to –5 and has no neutral zero point.

- Thurstone scale – This is a scaling technique that incorporates the intensity structure among indicators.
- Mathematically derived scale – Researchers infer respondents' evaluations mathematically. Two examples are multi dimensional scaling and conjoint analysis.

Scale Evaluation

Scales should be tested for reliability, generalizability, and validity. Generalizability is the ability to make inferences from a sample to the population, given the scale you have selected. Reliability is the extent to which a scale will produce consistent results. Test-retest reliability checks how similar the results are if the research is repeated under similar circumstances. Alternative forms reliability checks how similar the results are if the research is repeated using different forms of the scale. Internal consistency reliability checks how well the individual measures included in the scale are converted into a composite measure.

Scales and indexes have to be validated. Internal validation checks the relation between the individual measures included in the scale, and the composite scale itself. External validation checks the relation between the composite scale and other indicators of the variable, indicators not included in the scale. Content validation (also called face validity) checks how well the scale measures what is supposed to measured. Criterion validation checks how meaningful the scale criteria are relative to other possible criteria. Construct validation checks what underlying construct is being measured. There are three variants of construct validity. They are convergent validity, discriminant validity, and nomological validity. The coefficient of reproducibility indicates how well the data from the individual measures included in the scale can be reconstructed from the composite scale.

Logarithm

In mathematics, the logarithm of a number is the exponent to which another fixed value, the base, must be raised to produce that number. For example, the logarithm of 1000 to base 10 is 3, because 10 to the power 3 is 1000: $1000 = 10 \times 10 \times 10 = 10^3$. More generally, for any two real numbers b and x where b is positive and $b \neq 1$,

$$y = b^x \Leftrightarrow x = \log_b(y)$$

The logarithm to base 10 ($b = 10$) is called the common logarithm and has many applications in science and engineering. The natural logarithm has the irrational (transcendental) number e (≈ 2.718) as its base; its use is widespread in pure mathematics, especially calculus.

The binary logarithm uses base 2 ($b = 2$) and is prominent in computer science.

Logarithms were introduced by John Napier in the early 17th century as a means to simplify calculations. They were rapidly adopted by navigators, scientists, engineers, and others to perform computations more easily, using slide rules and logarithm tables. Tedious multi-digit multiplication steps can be replaced by table look-ups and simpler addition because of the fact—important in its own right—that the logarithm of a product is the sum of the logarithms of the factors:

$$\log_b(xy) = \log_b(x) + \log_b(y),$$

provided that b, x and y are all positive and $b \neq 1$. The present-day notion of logarithms comes from Leonhard Euler, who connected them to the exponential function in the 18th century.

Logarithmic scales reduce wide-ranging quantities to smaller scopes. For example, the decibel is a logarithmic unit quantifying signal power ratios and amplitude ratios (of which sound pressure is a common example). In chemistry, pH is a logarithmic measure for the acidity of an aqueous solution. Logarithms are commonplace in scientific formulae, and in measurements of the complexity of algorithms and of geometric objects called fractals. They describe musical intervals, appear in formulae counting prime numbers, inform some models in psychophysics, and can aid in forensic accounting. In the same way as the logarithm reverses exponentiation, the complex logarithm is the inverse function of the exponential function applied to complex numbers. The discrete logarithm is another variant; it has applications in public-key cryptography. The idea of logarithms is to reverse the operation of exponentiation, that is raising a number to a power. For example, the third power (or cube) of 2 is 8, because 8 is the product of three factors of 2:

$$2^3 = 2 \times 2 \times 2 = 8.$$

It follows that the logarithm of 8 with respect to base 2 is 3, so $\log_2 8 = 3$.

Exponentiation

The third power of some number b is the product of three factors of b. More generally, raising b to the n-th power, where n is a natural number, is done by multiplying n factors of b. The n-th power of b is written b^n, so that

$$b^n = \underbrace{b \times b \times \cdots \times b}_{n \text{ factors}}.$$

Exponentiation may be extended to b^y, where b is a positive number and the *exponent* y is any real number. For example, b^{-1} is the reciprocal of b, that is, $1/b$. (For further details, including the formula $b^{m+n} = b^m \cdot b^n$.).

Definition

The *logarithm* of a positive real number x with respect to base b, a positive real number not equal to 1[nb 1], is the exponent by which b must be raised to yield x. In other words, the logarithm of x to base b is the solution y to the equation

$$b^y = x.$$

The logarithm is denoted "$\log_b(x)$" (pronounced as "the logarithm of x to base b" or "the base-b logarithm of x"). In the equation $y = \log_b(x)$, the value y is the answer to the question "To what power must b be raised, in order to yield x?".

Examples

For example, $\log_2(16) = 4$, since $2^4 = 2 \times 2 \times 2 \times 2 = 16$. Logarithms can also be negative:

$$\log_2\left(\frac{1}{2}\right) = -1,$$

since

$$2^{-1} = \frac{1}{2^1} = \frac{1}{2}.$$

A third example: $\log_{10}(150)$ is approximately 2.176, which lies between 2 and 3, just as 150 lies between $10^2 = 100$ and $10^3 = 1000$. Finally, for any base b, $\log_b(b) = 1$ and $\log_b(1) = 0$, since $b^1 = b$ and $b^0 = 1$, respectively.

Logarithmic Identities

Several important formulas, sometimes called *logarithmic identities* or *log laws*, relate logarithms to one another.

Product, Quotient, Power and Root

The logarithm of a product is the sum of the logarithms of the numbers being multiplied; the logarithm of the ratio of two numbers is the difference of the logarithms. The logarithm of the p-th power of a number is p times the logarithm of the number itself; the logarithm of a p-th root is the logarithm of the number divided by p. The following table lists these identities with examples. Each of the identities can be derived after substitution of the logarithm definitions $x = b^{\log_b(x)}$, and/or $y = b^{\log_b(y)}$, in the left hand sides.

	Formula	*Example*
product	$\log_b(xy) = \log_b(x) + \log_b(y)$	$\log_3(243) = \log_3(9 \cdot 27) = \log_3(9) + \log_3(27) = 2 + 3 = 5$
quotient	$\log_b\left(\frac{x}{y}\right) = \log_b(x) - \log_b(y)$	$\log_2(16) = \log_2\left(\frac{64}{4}\right) = \log_2(64) - \log_2(4) = 6 - 2 = 4$
power	$\log_b(x^p) = p\log_b(x)$	$\log_2(64) = \log_2(2^6) = 6\log_2(2) = 6$
root	$\log_b \sqrt[p]{x} = \frac{\log_b(x)}{p}$	$\log_{10}\sqrt{1000} = \frac{1}{2}\log_{10}1000 = \frac{3}{2} = 1.5$

Change of Base

The logarithm $\log_b(x)$ can be computed from the logarithms of x and b with respect to an arbitrary base k using the following formula:

$$\log_b(x) = \frac{\log_k(x)}{\log_k(b)}.$$

Typical scientific calculators calculate the logarithms to bases 10 and e. Logarithms with respect to any base b can be determined using either of these two logarithms by the previous formula:

$$\log_b(x) = \frac{\log_{10}(x)}{\log_{10}(b)} = \frac{\log_e(x)}{\log_e(b)}.$$

Given a number x and its logarithm $\log_b(x)$ to an unknown base b, the base is given by:

$$b = x^{\frac{1}{\log_b(x)}}.$$

Particular Bases

Among all choices for the base, three are particularly common. These are $b = 10$, $b = e$ (the irrational mathematical constant ≈ 2.71828), and $b = 2$. In mathematical analysis, the logarithm to base e is widespread because of its particular analytical properties explained below. On the other hand, base-10 logarithms are easy to use for manual calculations in the decimal number system:

$$\log_{10}(10x) = \log_{10}(10) + \log_{10}(x) = 1 + \log_{10}(x).$$

Thus, $\log_{10}(x)$ is related to the number of decimal digits of a positive integer x: the number of digits is the smallest integer strictly bigger than $\log_{10}(x)$. For example, $\log_{10}(1430)$ is approximately 3.15. The next integer is 4, which is the number of digits of 1430. Both the natural logarithm and the logarithm to base two are used in information theory, corresponding to the use of nats or bits as the fundamental units of information, respectively. Binary logarithms are also used in computer science, where the binary system is ubiquitous, in music theory, where a pitch ratio of two (the octave) is ubiquitous and the cent is the binary logarithm (scaled by 1200) of the ratio between two adjacent equally-tempered pitches, and in photography to measure exposure values.

The following table lists common notations for logarithms to these bases and the fields where they are used. Many disciplines write log(x) instead of $\log_b(x)$, when the intended base can be determined from the context. The notation $^b\log(x)$ also occurs. The "ISO notation" column lists designations suggested by the International Organisation for Standardization (ISO 31-11).

Base b	***Name for $\log_b(x)$***	***ISO notation***	***Other notations***	***Used in***
2	binary logarithm	lb(x)	ld(x), log(x), lg(x), log2(x)	computer science, information theory, music theory, photography
e	natural logarithm	ln(x)	log(x) (in mathematics and many programming languages)	mathematics, physics, chemistry, statistics, economics, information theory, and some engineering fields
10	common logarithm	lg(x)	log(x), log10(x) (in engineering, biology, astronomy)	various engineering fields (see decibel and see below), logarithm tables, handheld calculators, spectroscopy

Logarithm Tables, Slide Rules, and Historical Applications

By simplifying difficult calculations, logarithms contributed to the advance of science, and especially of astronomy. They were critical to advances in surveying, celestial navigation, and other domains. Pierre-Simon Laplace called logarithms

> *"...[a]n admirable artifice which, by reducing to a few days the labour of many months, doubles the life of the astronomer, and spares him the errors and disgust inseparable from long calculations."*

A key tool that enabled the practical use of logarithms before calculators and computers was the *table of logarithms*. The first such table was compiled by Henry Briggs in 1617, immediately after Napier's invention. Subsequently, tables with increasing scope and precision were written. These tables listed the values of $\log_b(x)$ and b^x for any number x in a certain range, at a certain precision, for a certain base b (usually $b = 10$). For example, Briggs' first table contained the common logarithms of all integers in the range 1–1000, with a precision of 8 digits. As the function $f(x) = b^x$ is the inverse function of $\log_b(x)$, it has been called the antilogarithm. The product and quotient of two positive numbers *c* and *d* were routinely calculated as the sum and difference of their logarithms. The product *cd* or quotient *c*/*d* came from looking up the antilogarithm of the sum or difference, also via the same table:

$$cd = b^{\log_b(c)} b^{\log_b(d)} = b^{\log_b(c)+\log_b(d)}$$

and

$$\frac{c}{d} = cd^{-1} = b^{\log_b(c)-\log_b(d)}.$$

For manual calculations that demand any appreciable precision, performing the lookups of the two logarithms, calculating their sum or difference, and looking up the antilogarithm is much faster than performing the multiplication by earlier methods such as prosthaphaeresis, which relies on trigonometric identities. Calculations of powers and roots are reduced to multiplications or divisions and look-ups by

$$c^d = (b^{\log_b(c)})^d = b^{d\log_b(c)}$$

and

$$\sqrt[d]{c} = c^{\frac{1}{d}} = b^{\frac{1}{d}\log_b(c)}.$$

Many logarithm tables give logarithms by separately providing the characteristic and mantissa of *x*, that is to say, the integer part and the fractional part of $\log_{10}(x)$. The characteristic of $10 \cdot x$ is one plus the characteristic of *x*, and their significants are the same. This extends the scope of logarithm tables: given a table listing $\log_{10}(x)$ for all integers *x* ranging from 1 to 1000, the logarithm of 3542 is approximated by

$$\log_{10}(3542) = \log_{10}(10 \cdot 354.2) = 1 + \log_{10}(354.2) \approx 1 + \log_{10}(354).$$

Another critical application was the slide rule, a pair of logarithmically divided scales used for calculation, as illustrated here:

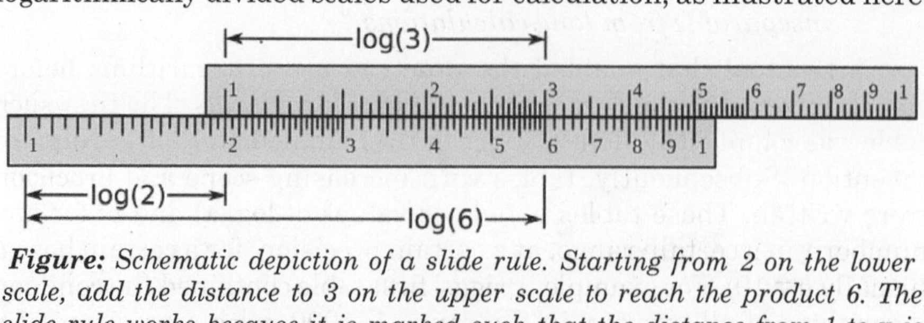

Figure: *Schematic depiction of a slide rule. Starting from 2 on the lower scale, add the distance to 3 on the upper scale to reach the product 6. The slide rule works because it is marked such that the distance from 1 to* x *is proportional to the logarithm of* x.

The non-sliding logarithmic scale, Gunter's rule, was invented shortly after Napier's invention. William Oughtred enhanced it to create the slide rule—a pair of logarithmic scales movable with respect to each other. Numbers are placed on sliding scales at distances proportional to the differences between their logarithms. Sliding the upper scale appropriately amounts to mechanically adding logarithms. For example, adding the distance from 1 to 2 on the lower scale to the distance from 1 to 3 on the upper scale yields a product of 6, which is read off at the lower part. The slide rule was an essential calculating tool for engineers and scientists until the 1970s, because it allows, at the expense of precision, much faster computation than techniques based on tables.

Analytic Properties

A deeper study of logarithms requires the concept of a function. A function is a rule that, given one number, produces another number. An example is the function producing the x-th power of b from any real number x, where the base b is a fixed number. This function is written

$$f(x) = b^x.$$

Logarithmic Function

To justify the definition of logarithms, it is necessary to show that the equation

$$b^x = y$$

has a solution x and that this solution is unique, provided that y is positive and that b is positive and unequal to 1. A proof of that fact requires the intermediate value theorem from elementary calculus. This theorem states that a continuous function that produces two values m and n also produces any value that lies between m and n. A function is *continuous* if it does not "jump", that is, if its graph can be drawn without lifting the pen.

This property can be shown to hold for the function $f(x) = b^x$. Because f takes arbitrarily large and arbitrarily small positive values, any number $y > 0$ lies between $f(x_0)$ and $f(x_1)$ for suitable x_0 and x_1. Hence, the intermediate value theorem ensures that the equation $f(x) = y$ has a solution. Moreover, there is only one solution to this equation, because the function f is strictly increasing (for $b > 1$), or strictly decreasing (for $0 < b < 1$).

The unique solution x is the logarithm of y to base b, $\log_b(y)$. The function that assigns to y its logarithm is called *logarithm function* or *logarithmic function* (or just *logarithm*).

The function $\log_b(x)$ is essentially characterized by the above product formula

$$\log_b(xy) = \log_b(x) + \log_b(y).$$

More precisely, the logarithm to any base $b > 1$ is the only increasing function f from the positive reals to the reals satisfying $f(b) = 1$ and

$$f(xy) = f(x) + f(y).$$

Inverse Function

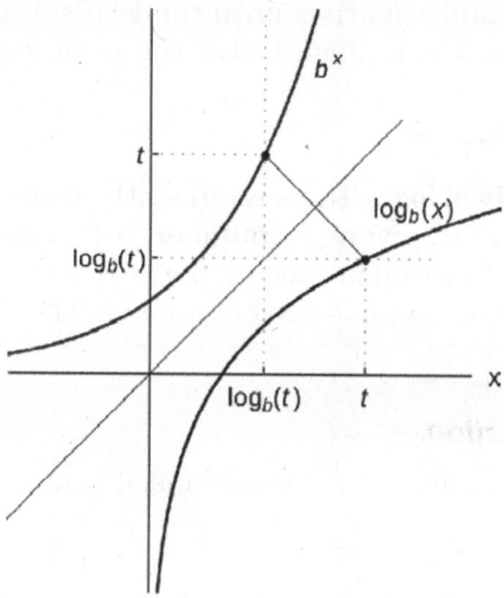

***Figure:** The graph of the logarithm function* $\log_b(x)$ *(blue) is obtained by reflecting the graph of the function* b^x *(red) at the diagonal line* $(x = y)$.

The formula for the logarithm of a power says in particular that for any number x,

$$\log_b\left(b^x\right) = x\log_b(b) = x.$$

In prose, taking the x-th power of b and then the base-b logarithm gives back x. Conversely, given a positive number y, the formula

$$b^{\log_b(y)} = y$$

says that first taking the logarithm and then exponentiating gives back y. Thus, the two possible ways of combining (or composing) logarithms and exponentiation give back the original number. Therefore, the logarithm to base b is the inverse function of $f(x) = b^x$.

Inverse functions are closely related to the original functions. Their graphs correspond to each other upon exchanging the x- and the y-coordinates (or upon reflection at the diagonal line $x = y$), as shown at the right: a point $(t, u = b^t)$ on the graph of f yields a point $(u, t = \log_b u)$ on the graph of the logarithm and vice versa. As a consequence, $\log_b(x)$ diverges to infinity (gets bigger than any given number) if x grows to infinity, provided that b is greater than one. In that case, $\log_b(x)$ is an increasing function. For $b < 1$, $\log_b(x)$ tends to minus infinity instead. When x approaches zero, $\log_b(x)$ goes to minus infinity for $b > 1$ (plus infinity for $b < 1$, respectively).

Derivative and Antiderivative

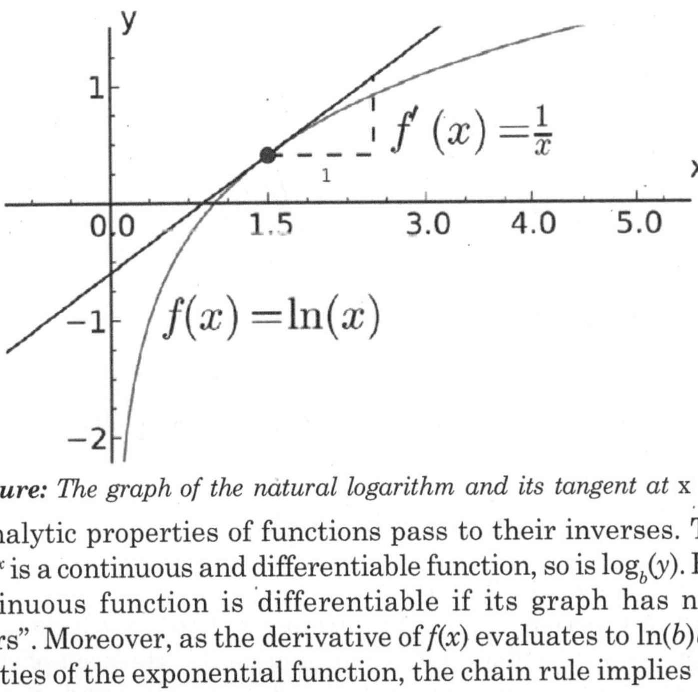

Figure: *The graph of the natural logarithm and its tangent at* x = *1.5*

Analytic properties of functions pass to their inverses. Thus, as $f(x) = b^x$ is a continuous and differentiable function, so is $\log_b(y)$. Roughly, a continuous function is differentiable if its graph has no sharp "corners". Moreover, as the derivative of $f(x)$ evaluates to $\ln(b)b^x$ by the properties of the exponential function, the chain rule implies that the derivative of $\log_b(x)$ is given by

$$\frac{d}{dx}\log_b(x) = \frac{1}{x\ln(b)}.$$

That is, the slope of the tangent touching the graph of the base-b logarithm at the point $(x, \log_b(x))$ equals $1/(x \ln(b))$. In particular, the derivative of $\ln(x)$ is $1/x$, which implies that the antiderivative of $1/x$ is $\ln(x) + C$. The derivative with a generalised functional argument $f(x)$ is

$$\frac{d}{dx}\ln(f(x)) = \frac{f'(x)}{f(x)}.$$

The quotient at the right hand side is called the logarithmic derivative of f. Computing $f'(x)$ by means of the derivative of $\ln(f(x))$ is known as logarithmic differentiation. The antiderivative of the natural logarithm ln(*x*) is:

$$\int \ln(x)dx = x\ln(x) - x + C.$$

Related formulas, such as antiderivatives of logarithms to other bases can be derived from this equation using the change of bases.

Integral Representation of the Natural Logarithm

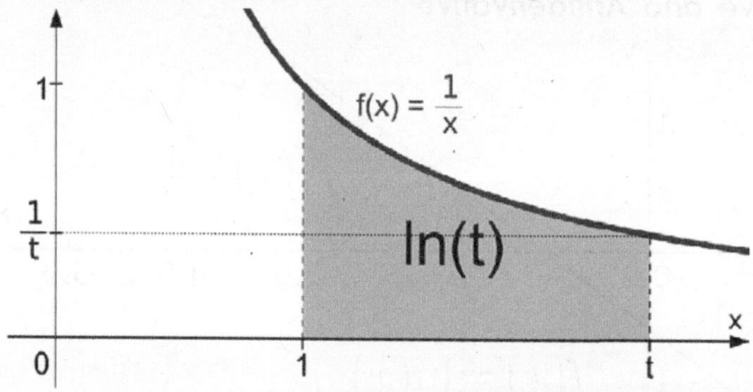

***Figure:** The natural logarithm of* t *is the shaded area underneath the graph of the function* f(x) = *1/*x *(reciprocal of* x*).*

The natural logarithm of t agrees with the integral of $1/x\ dx$ from 1 to t:

$$\ln(t) = \int_1^t \frac{1}{x}dx.$$

In other words, ln(t) equals the area between the x axis and the graph of the function $1/x$, ranging from $x = 1$ to $x = t$ (figure at the right). This is a consequence of the fundamental theorem of calculus and the fact that derivative of ln(x) is $1/x$. The right hand side of this equation can serve as a definition of the natural logarithm. Product and power logarithm formulas can be derived from this definition. For example, the product formula $\ln(tu) = \ln(t) + \ln(u)$ is deduced as:

$$\ln(tu) = \int_1^{tu} \frac{1}{x}dx \overset{(1)}{=} \int_1^t \frac{1}{x}dx + \int_t^{tu} \frac{1}{x}dx \overset{(2)}{=} \ln(t) + \int_1^u \frac{1}{w}dw = \ln(t) + \ln(u).$$

The equality (1) splits the integral into two parts, while the equality (2) is a change of variable ($w = x/t$). In the illustration below, the splitting corresponds to dividing the area into the yellow and blue parts. Rescaling the left hand blue area vertically by the factor t and shrinking it by the same factor horizontally does not change its size. Moving it appropriately, the area fits the graph of the function $f(x) = 1/x$ again. Therefore, the left hand blue area, which is the integral of $f(x)$ from t to

tu is the same as the integral from 1 to u. This justifies the equality (2) with a more geometric proof.

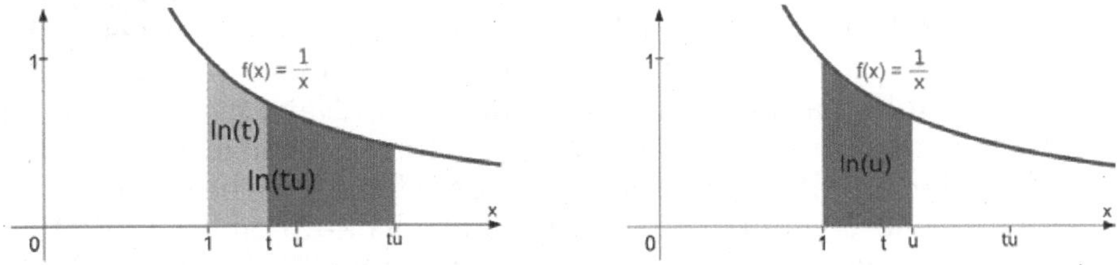

***Figure:** A visual proof of the product formula of the natural logarithm*

The power formula $\ln(t^r) = r\ln(t)$ may be derived in a similar way:

$$\ln(t^r) = \int_1^{t^r} \frac{1}{x}dx = \int_1^t \frac{1}{w^r}\left(rw^{r-1}\,dw\right) = r\int_1^t \frac{1}{w}dw = r\ln(t).$$

The second equality uses a change of variables (integration by substitution), $w = x^{1/r}$.

The sum over the reciprocals of natural numbers,

$$1+\frac{1}{2}+\frac{1}{3}+\cdots+\frac{1}{n}=\sum_{k=1}^{n}\frac{1}{k},$$

is called the harmonic series. It is closely tied to the natural logarithm: as n tends to infinity, the difference,

$$\sum_{k=1}^{n}\frac{1}{k}-\ln(n),$$

converges (i.e., gets arbitrarily close) to a number known as the Euler–Mascheroni constant. This relation aids in analyzing the performance of algorithms such as quicksort.

There is also another integral representation of the logarithm that is useful in some situations.

$$\ln(x) = -\lim_{\epsilon\to 0}\int_\epsilon^\infty \frac{dt}{t}\left(e^{-xt}-e^{-t}\right)$$

This can be verified by showing that it has the same value at $x = 1$, and the same derivative.

Transcendence of the Logarithm

Real numbers that are not algebraic are called transcendental; for example, π and e are such numbers, but $\sqrt{2-\sqrt{3}}$ is not. Almost all real numbers are transcendental. The logarithm is an example of a transcendental function. The Gelfond–Schneider theorem asserts that logarithms usually take transcendental, i.e., "difficult" values.

Calculation

Logarithms are easy to compute in some cases, such as $\log_{10}(1{,}000) = 3$. In general, logarithms can be calculated using power series or the arithmetic-geometric mean, or be retrieved from a precalculated logarithm table that provides a fixed precision. Newton's method, an iterative method to solve equations approximately, can also be used to calculate the logarithm, because its inverse function, the exponential function, can be computed efficiently. Using look-up tables, CORDIC-like methods can be used to compute logarithms if the only available operations are addition and bit shifts. Moreover, the binary logarithm algorithm calculates lb(x) recursively based on repeated squarings of x, taking advantage of the relation

$$\log_2(x^2) = 2\log_2(x).$$

Power Series

Taylor series:

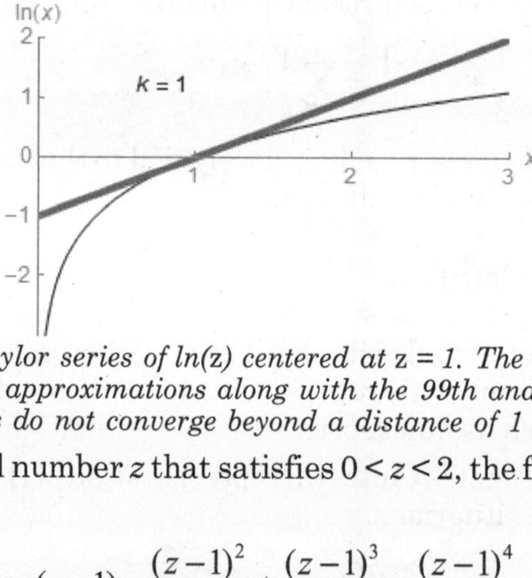

***Figure"** The Taylor series of ln(z) centered at z = 1. The animation shows the first 10 approximations along with the 99th and 100th. The approximations do not converge beyond a distance of 1 from the center.*

For any real number z that satisfies $0 < z < 2$, the following formula holds: [nb 4]

$$\ln(z) = (z-1) - \frac{(z-1)^2}{2} + \frac{(z-1)^3}{3} - \frac{(z-1)^4}{4} + \cdots$$

This is a shorthand for saying that ln(z) can be approximated to a more and more accurate value by the following expressions:

$$(z-1)$$

$$(z-1) \quad - \quad \frac{(z-1)^2}{2}$$

$$(z-1) \quad - \quad \frac{(z-1)^2}{2} \quad + \quad \frac{(z-1)^3}{3}$$

$$\vdots$$

For example, with $z = 1.5$ the third approximation yields 0.4167, which is about 0.011 greater than ln(1.5) = 0.405465. This series approximates ln(z) with arbitrary precision, provided the number of summands is large enough. In elementary calculus, ln(z) is therefore the limit of this series. It is the Taylor series of the natural logarithm at z = 1. The Taylor series of ln z provides a particularly useful approximation to ln(1+z) when z is small, $|z| < 1$, since then

$$\ln(1+z) = z - \frac{z^2}{2} + \frac{z^3}{3} \cdots \approx z.$$

For example, with $z = 0.1$ the first-order approximation gives ln(1.1) H" 0.1, which is less than 5% off the correct value 0.0953.

More Efficient Series

Another series is based on the area hyperbolic tangent function:

$$\ln(z) = 2\cdot\text{artanh}\frac{z-1}{z+1} = 2\left(\frac{z-1}{z+1} + \frac{1}{3}\left(\frac{z-1}{z+1}\right)^3 + \frac{1}{5}\left(\frac{z-1}{z+1}\right)^5 + \cdots\right),$$

for any real number $z > 0$.[nb 5] Using the Sigma notation, this is also written as

$$\ln(z) = 2\sum_{n=0}^{\infty} \frac{1}{2n+1}\left(\frac{z-1}{z+1}\right)^{2n+1}.$$

This series can be derived from the above Taylor series. It converges more quickly than the Taylor series, especially if z is close to 1. For example, for z = 1.5, the first three terms of the second series approximate ln(1.5) with an error of about 3×10^{-6}. The quick convergence for z close to 1 can be taken advantage of in the following way: given a low-accuracy approximation $y \approx \ln(z)$ and putting

$$A = \frac{z}{\exp(y)},$$

the logarithm of z is:

$$\ln(z) = y + \ln(A).$$

The better the initial approximation y is, the closer A is to 1, so its logarithm can be calculated efficiently. A can be calculated using the exponential series, which converges quickly provided y is not too large. Calculating the logarithm of larger z can be reduced to smaller values of z by writing $z = a \cdot 10^b$, so that $\ln(z) = \ln(a) + b \cdot \ln(10)$.

A closely related method can be used to compute the logarithm of integers. From the above series, it follows that:

$$\ln(n+1) = \ln(n) + 2\sum_{k=0}^{\infty} \frac{1}{2k+1}\left(\frac{1}{2n+1}\right)^{2k+1}.$$

If the logarithm of a large integer n is known, then this series yields a fast converging series for $\log(n+1)$.

Arithmetic-Geometric Mean Approximation

The arithmetic-geometric mean yields high precision approximations of the natural logarithm. $\ln(x)$ is approximated to a precision of 2^{-p} (or p precise bits) by the following formula (due to Carl Friedrich Gauss):

$$\ln(x) \approx \frac{\pi}{2M(1, 2^{2-m}/x)} - m\ln(2).$$

Here M denotes the arithmetic-geometric mean. It is obtained by repeatedly calculating the average (arithmetic mean) and the square root of the product of two numbers (geometric mean). Moreover, m is chosen such that

$$x2^m > 2^{p/2}.$$

Both the arithmetic-geometric mean and the constants π and $\ln(2)$ can be calculated with quickly converging series.

Applications

Logarithms have many applications inside and outside mathematics. Some of these occurrences are related to the notion of scale invariance. For example, each chamber of the shell of a nautilus is an approximate copy of the next one, scaled by a constant factor. This gives rise to a logarithmic spiral. Benford's law on the distribution of leading digits can also be explained by scale invariance. Logarithms are also linked to self-similarity.

For example, logarithms appear in the analysis of algorithms that solve a problem by dividing it into two similar smaller problems and patching their solutions. The dimensions of self-similar geometric shapes, that is, shapes whose parts resemble the overall picture are also based on logarithms. Logarithmic scales are useful for quantifying the relative change of a value as opposed to its absolute difference. Moreover, because the logarithmic function $\log(x)$ grows very slowly for large x, logarithmic scales are used to compress large-scale scientific data. Logarithms also occur in numerous scientific formulas, such as the Tsiolkovsky rocket equation, the Fenske equation, or the Nernst equation.

Logarithmic Scale

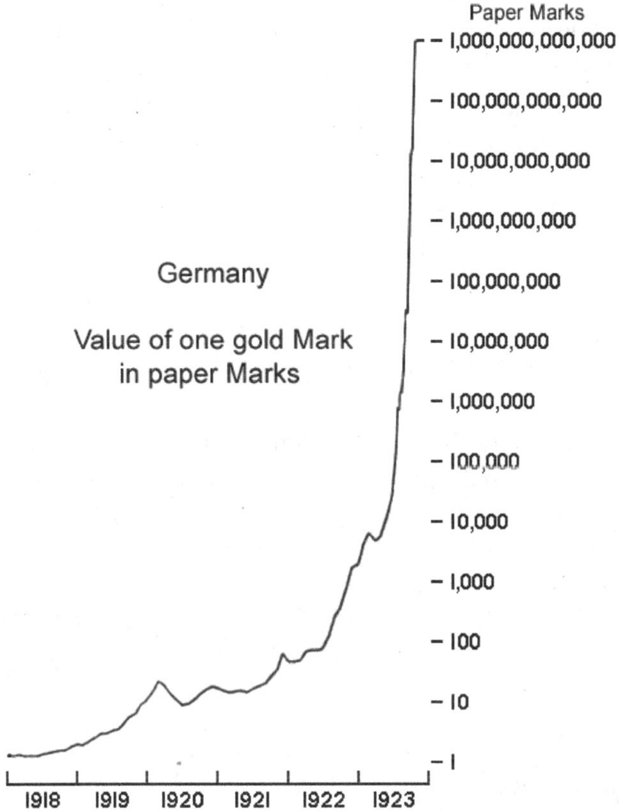

Fig. A logarithmic chart depicting the value of one Goldmark in Papiermarks during the German hyperinflation in the 1920s

Scientific quantities are often expressed as logarithms of other quantities, using a *logarithmic scale.* For example, the decibel is a logarithmic unit of measurement. It is based on the common logarithm of ratios—10 times the common logarithm of a power ratio or 20 times the common logarithm of a voltage ratio. It is used to quantify the loss of voltage levels in transmitting electrical signals, to describe power levels of sounds in acoustics, and the absorbance of light in the fields of spectrometry and optics. The signal-to-noise ratio describing the amount of unwanted noise in relation to a (meaningful) signal is also measured in decibels. In a similar vein, the peak signal-to-noise ratio is commonly used to assess the quality of sound and image compression methods using the logarithm.

The strength of an earthquake is measured by taking the common logarithm of the energy emitted at the quake. This is used in the moment magnitude scale or the Richter scale. For example, a 5.0 earthquake releases 10 times and a 6.0 releases 100 times the energy of a 4.0. Another logarithmic scale is apparent magnitude. It measures the brightness of stars logarithmically.

Yet another example is pH in chemistry; pH is the negative of the common logarithm of the activity of hydronium ions (the form hydrogen ions H^+ take in water). The activity of hydronium ions in neutral water is 10^{-7} mol·L^{-1}, hence a pH of 7. Vinegar typically has a pH of about 3. The difference of 4 corresponds to a ratio of 10^4 of the activity, that is, vinegar's hydronium ion activity is about 10^{-3} mol·L^{-1}.

Semilog (log-linear) graphs use the logarithmic scale concept for visualization: one axis, typically the vertical one, is scaled logarithmically.

For example, the chart at the right compresses the steep increase from 1 million to 1 trillion to the same space (on the vertical axis) as the increase from 1 to 1 million. In such graphs, exponential functions of the form $f(x) = a \cdot b^x$ appear as straight lines with slope equal to the logarithm of b. Log-log graphs scale both axes logarithmically, which causes functions of the form $f(x) = a \cdot x^k$ to be depicted as straight lines with slope equal to the exponent k. This is applied in visualizing and analyzing power laws.

Psychology

Logarithms occur in several laws describing human perception: Hick's law proposes a logarithmic relation between the time individuals take for choosing an alternative and the number of choices they have. Fitts's law predicts that the time required to rapidly move to a target area is a logarithmic function of the distance to and the size of the target. In psychophysics, the Weber–Fechner law proposes a logarithmic relationship between stimulus and sensation such as the actual vs. the perceived weight of an item a person is carrying. (This "law", however, is less precise than more recent models, such as the Stevens' power law.)

Psychological studies found that individuals with little mathematics education tend to estimate quantities logarithmically, that is, they position a number on an unmarked line according to its logarithm, so that 10 is positioned as close to 100 as 100 is to 1000. Increasing education shifts this to a linear estimate (positioning 1000 10x as far away) in some circumstances, while logarithms are used when the numbers to be plotted are difficult to plot linearly.

Probability Theory and Statistics

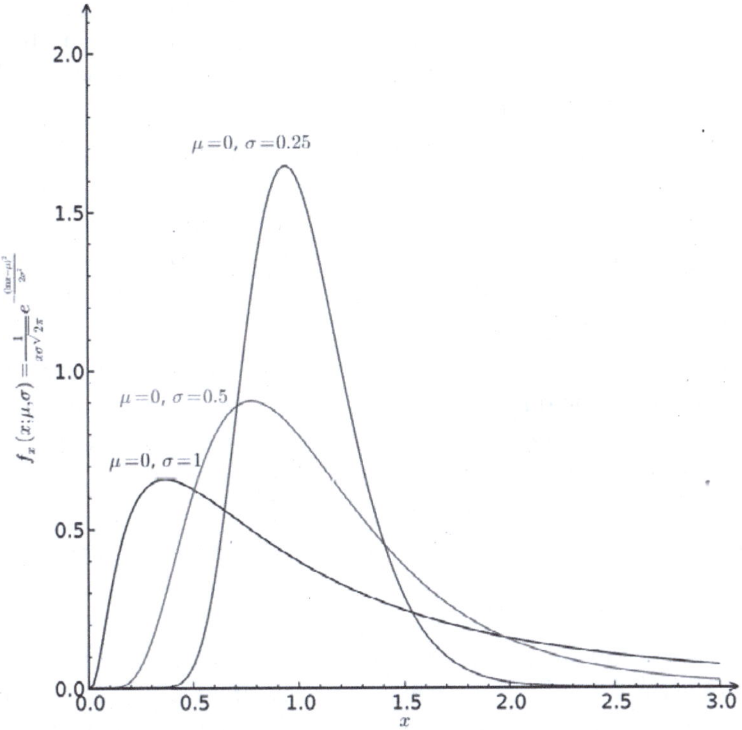

***Figure:** Three probability density functions of random variables with log-normal distributions. The location parameter μ, which is zero for all three of the PDFs shown, is the mean of the logarithm of the random variable, not the mean of the variable itself.*

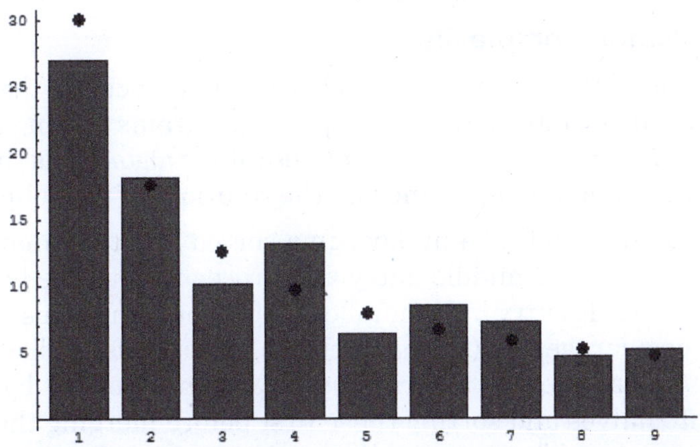

***Figure:** Distribution of first digits (in %, red bars) in the population of the 237 countries of the world. Black dots indicate the distribution predicted by Benford's law.*

Logarithms arise in probability theory: the law of large numbers dictates that, for a fair coin, as the number of coin-tosses increases to infinity, the observed proportion of heads approaches one-half. The fluctuations of this proportion about one-half are described by the law of the iterated logarithm.

Logarithms also occur in log-normal distributions. When the logarithm of a random variable has a normal distribution, the variable is said to have a log-normal distribution. Log-normal distributions are encountered in many fields, wherever a variable is formed as the product of many independent positive random variables, for example in the study of turbulence.

Logarithms are used for maximum-likelihood estimation of parametric statistical models. For such a model, the likelihood function depends on at least one parameter that must be estimated. A maximum of the likelihood function occurs at the same parameter-value as a maximum of the logarithm of the likelihood (the "*log likelihood*"), because the logarithm is an increasing function. The log-likelihood is easier to maximize, especially for the multiplied likelihoods for independent random variables.

Benford's law describes the occurrence of digits in many data sets, such as heights of buildings. According to Benford's law, the probability that the first decimal-digit of an item in the data sample is d (from 1 to 9) equals $\log_{10}(d + 1) - \log_{10}(d)$, *regardless* of the unit of measurement. Thus, about 30% of the data can be expected to have 1 as first digit, 18% start with 2, etc. Auditors examine deviations from Benford's law to detect fraudulent accounting.

Computational Complexity

Analysis of algorithms is a branch of computer science that studies the performance of algorithms (computer programs solving a certain problem). Logarithms are valuable for describing algorithms that divide a problem into smaller ones, and join the solutions of the subproblems.

For example, to find a number in a sorted list, the binary search algorithm checks the middle entry and proceeds with the half before or after the middle entry if the number is still not found. This algorithm requires, on average, $\log_2(N)$ comparisons, where N is the list's length. Similarly, the merge sort algorithm sorts an unsorted list by dividing the list into halves and sorting these first before merging the results. Merge sort algorithms typically require a time approximately proportional to $N \cdot \log(N)$. The base of the logarithm is not specified here, because the result only changes by a constant factor when another base is used. A constant factor, is usually disregarded in the analysis of algorithms under the standard uniform cost model.

A function $f(x)$ is said to grow logarithmically if $f(x)$ is (exactly or approximately) proportional to the logarithm of x. (Biological descriptions of organism growth, however, use this term for an exponential function.) For example, any natural number N can be represented in binary form in no more than $\log_2(N) + 1$ bits. In other words, the amount of memory needed to store N grows logarithmically with N.

Entropy and Chaos

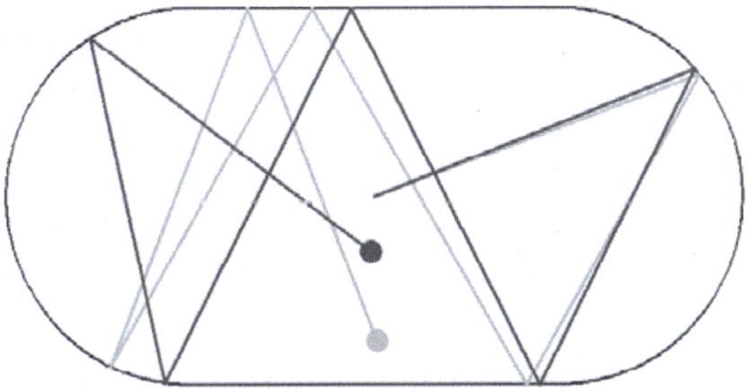

Figure: Billiards on an oval billiard table. Two particles, starting at the center with an angle differing by one degree, take paths that diverge chaotically because of reflections at the boundary.

Entropy is broadly a measure of the disorder of some system. In statistical thermodynamics, the entropy S of some physical system is defined as

$$S = -k\sum_i p_i \ln(p_i).$$

The sum is over all possible states i of the system in question, such as the positions of gas particles in a container. Moreover, p_i is the probability that the state i is attained and k is the Boltzmann constant. Similarly, entropy in information theory measures the quantity of information. If a message recipient may expect any one of N possible messages with equal likelihood, then the amount of information conveyed by any one such message is quantified as $\log_2(N)$ bits.

Lyapunov exponents use logarithms to gauge the degree of chaoticity of a dynamical system. For example, for a particle moving on an oval billiard table, even small changes of the initial conditions result in very different paths of the particle. Such systems are chaotic in a deterministic way, because small measurement errors of the initial state predictably lead to largely different final states. At least one Lyapunov exponent of a deterministically chaotic system is positive.

Fractals

***Figure:** The Sierpinski triangle (at the right) is constructed by repeatedly replacing equilateral triangles by three smaller ones.*

Logarithms occur in definitions of the dimension of fractals. Fractals are geometric objects that are self-similar: small parts reproduce, at least roughly, the entire global structure. The Sierpinski triangle (pictured) can be covered by three copies of itself, each having sides half the original length. This makes the Hausdorff dimension of this structure log(3)/log(2) H" 1.58. Another logarithm-based notion of dimension is obtained by counting the number of boxes needed to cover the fractal in question.

Music

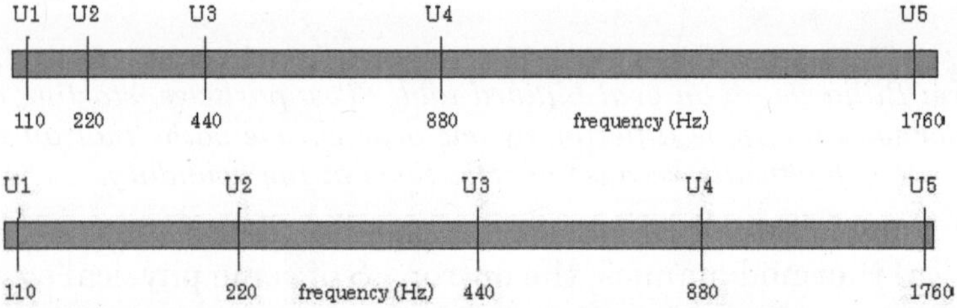

***Figure:** Four different octaves shown on a linear scale, then shown on a logarithmic scale (as the ear hears them).*

Logarithms are related to musical tones and intervals. In equal temperament, the frequency ratio depends only on the interval between two tones, not on the specific frequency, or pitch, of the individual tones. For example, the note A has a frequency of 440 Hz and B-flat has a frequency of 466 Hz. The interval between *A* and *B-flat* is a semitone, as is the one between *B-flat* and B (frequency 493 Hz). Accordingly, the frequency ratios agree:

$$\frac{466}{440} \approx \frac{493}{466} \approx 1.059 \approx \sqrt[12]{2}.$$

Therefore, logarithms can be used to describe the intervals: an interval is measured in semitones by taking the base-$2^{1/12}$ logarithm of the frequency ratio, while the base-$2^{1/1200}$ logarithm of the frequency ratio expresses the interval in cents, hundredths of a semitone. The latter is used for finer encoding, as it is needed for non-equal temperaments.

Number Theory

Natural logarithms are closely linked to counting prime numbers (2, 3, 5, 7, 11, ...), an important topic in number theory. For any integer x, the quantity of prime numbers less than or equal to x is denoted $\pi(x)$. The prime number theorem asserts that $\pi(x)$ is approximately given by

$$\frac{x}{\ln(x)},$$

in the sense that the ratio of $\pi(x)$ and that fraction approaches 1 when x tends to infinity. As a consequence, the probability that a randomly chosen number between 1 and x is prime is inversely proportional to the numbers of decimal digits of x. A far better estimate of $\pi(x)$ is given by the offset logarithmic integral function Li(x), defined by

$$\mathrm{Li}(x) = \int_2^x \frac{1}{\ln(t)} dt.$$

The Riemann hypothesis, one of the oldest open mathematical conjectures, can be stated in terms of comparing $\pi(x)$ and Li(x). The Erdõs–Kac theorem describing the number of distinct prime factors also involves the natural logarithm.

The logarithm of n factorial, $n! = 1 \cdot 2 \cdot \ldots \cdot n$, is given by

$$\ln(n!) = \ln(1) + \ln(2) + \cdots + \ln(n).$$

This can be used to obtain Stirling's formula, an approximation of $n!$ for large n.

Generalizations

Complex logarithm:

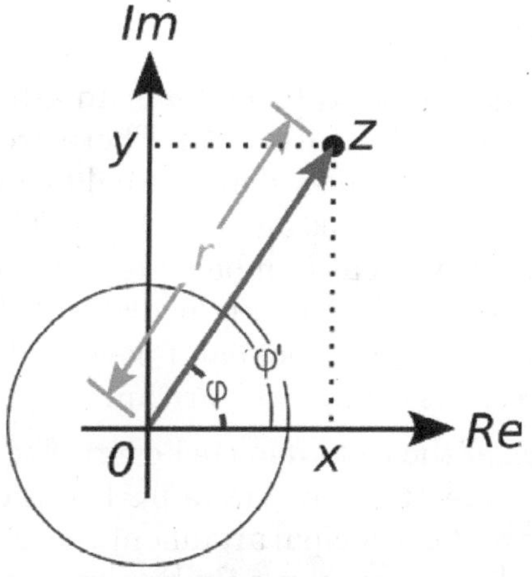

Figure: *Polar form of* z = x + iy. *Both* φ *and* φ' *are arguments of* z.

The complex numbers a solving the equation

$$e^a = z.$$

are called *complex logarithms*. Here, z is a complex number. A complex number is commonly represented as $z = x + iy$, where x and y are real numbers and i is the imaginary unit. Such a number can be visualized by a point in the complex plane, as shown at the right. The polar form encodes a non-zero complex number z by its absolute value, that is, the distance r to the origin, and an angle between the x axis and the line passing through the origin and z. This angle is called the argument of z. The absolute value r of z is

$$r = \sqrt{x^2 + y^2}.$$

The argument is not uniquely specified by z: both φ and $\varphi' = \varphi + 2\pi$ are arguments of z because adding 2π radians or 360 degrees[nb 6] to φ corresponds to "winding" around the origin counter-clock-wise by a turn. The resulting complex number is again z, as illustrated at the right. However, exactly one argument φ satisfies $-\pi < \varphi$ and $\varphi \leq \pi$. It is called the *principal argument*, denoted Arg(z), with a capital A. (An alternative normalization is $0 \leq \text{Arg}(z) < 2\pi$.)

Using trigonometric functions sine and cosine, or the complex exponential, respectively, r and φ are such that the following identities hold:

$$\begin{aligned} z &= r(\cos\varphi + i\sin\varphi) \\ &= re^{i\varphi}. \end{aligned}$$

This implies that the a-th power of e equals z, where

$$a = \ln(r) + i(\varphi + 2n\pi),$$

φ is the principal argument Arg(z) and n is an arbitrary integer. Any such a is called a complex logarithm of z. There are infinitely many of them, in contrast to the uniquely defined real logarithm. If $n = 0$, a is called the *principal value* of the logarithm, denoted Log(z). The principal argument of any positive real number x is 0; hence Log(x) is a real number and equals the real (natural) logarithm. However, the above formulas for logarithms of products and powers do not generalize to the principal value of the complex logarithm.

The illustration at the right depicts Log(z). The discontinuity, that is, the jump in the hue at the negative part of the x- or real axis, is caused by the jump of the principal argument there. This locus is called a branch cut. This behaviour can only be circumvented by dropping the range restriction on φ. Then the argument of z and, consequently, its logarithm become multi-valued functions.

Inverses of Other Exponential Functions

Exponentiation occurs in many areas of mathematics and its inverse function is often referred to as the logarithm. For example, the logarithm of a matrix is the (multi-valued) inverse function of the matrix exponential. Another example is the p-adic logarithm, the inverse function of the p-adic exponential. Both are defined via Taylor series analogous to the real case. In the context of differential geometry, the exponential map maps the tangent space at a point of a manifold to a neighbourhood of that point. Its inverse is also called the logarithmic (or log) map.

In the context of finite groups exponentiation is given by repeatedly multiplying one group element b with itself. The discrete logarithm is the integer n solving the equation

$$b^n = x,$$

where x is an element of the group. Carrying out the exponentiation can be done efficiently, but the discrete logarithm is believed to be very hard to calculate in some groups. This asymmetry has important applications in public key cryptography, such as for example in the Diffie–Hellman key exchange, a routine that allows secure exchanges of cryptographic keys over unsecured information channels. Zech's logarithm is related to the discrete logarithm in the multiplicative group of non-zero elements of a finite field.

Further logarithm-like inverse functions include the *double logarithm* $\ln(\ln(x))$, the super- or hyper-4-logarithm (a slight variation of which is called iterated logarithm in computer science), the Lambert W function, and the logit. They are the inverse functions of the double exponential function, tetration, of $f(w) = we^w$, and of the logistic function, respectively.

Related Concepts

From the perspective of pure mathematics, the identity $\log(cd) = \log(c) + \log(d)$ expresses a group isomorphism between positive reals under multiplication and reals under addition. Logarithmic functions are the only continuous isomorphisms between these groups. By means of that isomorphism, the Haar measure (Lebesgue measure) dx on the reals corresponds to the Haar measure dx/x on the positive reals. In complex analysis and algebraic geometry, differential forms of the form df/f are known as forms with logarithmic poles.

The polylogarithm is the function defined by

$$\mathrm{Li}_s(z) = \sum_{k=1}^{\infty} \frac{z^k}{k^s}.$$

It is related to the natural logarithm by $\mathrm{Li}_1(z) = -\ln(1 - z)$. Moreover, $\mathrm{Li}_s(1)$ equals the Riemann zeta function $\zeta(s)$.

Equations

In mathematics, an equation is a formula of the form $A = B$, where A and B are expressions that may contain one or several variables called unknowns, and "=" denotes the equality binary relation. Although written in the form of proposition, an equation is not a statement that is either true or false, but a problem consisting of finding the values, called solutions, that, when substituted for the unknowns, yield equal values of the expressions A and B. For example, 2 is the unique *solution* of the *equation* $x + 2 = 4$, in which the *unknown* is x. Historically, equations arose from the mathematical discipline of algebra, but later became ubiquitous. "Equations" should not be confused with "identities", which are presented with the same notation but have a different meaning: for example $2 + 2 = 4$ and $x + y = y + x$ are identities (which implies they are necessarily true) in arithmetic, and do not constitute a values-finding problem, even when variables are present as in the latter example. The term "equation" may also refer to a relation between some variables that is presented as the equality of some expressions written in terms of those variables' values. For example the *equation* of the unit circle is $x^2 + y^2 = 1$, which means that a point belongs to the circle if and only if its coordinates are related by this equation. Most physical laws are expressed by equations. One of the most famous ones is Einstein's equation E = mc2.

The = symbol was invented by Robert Recorde (1510–1558), who considered that nothing could be more equal than parallel straight lines with the same length.

Parameters and Unknowns

Equations often contain terms other than the unknowns. These other terms, which are assumed to be *known*, are usually called *constants*, *coefficients* or *parameters*. Usually, the unknowns are denoted by letters at the end of the alphabet, $x, y, z, w, \ldots$, while coefficients are denoted by letters at the beginning, $a, b, c, d, \ldots$. For example, the general quadratic equation is usually written $ax^2 + bx + c = 0$. The process of finding the solutions, or in case of parameters, expressing the unknowns in terms of the parameters is called solving the equation. Such expressions of the solutions in terms of the parameters are also called *solutions*.

A system of equations is a set of *simultaneous equations*, usually in several unknowns, for which the common solutions are sought. Thus a *solution to the system* is a set of values for each of the unknowns, which together form a solution to each equation in the system. For example, the system

$$3x+5y = 2$$
$$5x+8y = 3$$

has the unique solution $x =$ •1, $y = 1$.

Analogous Illustration

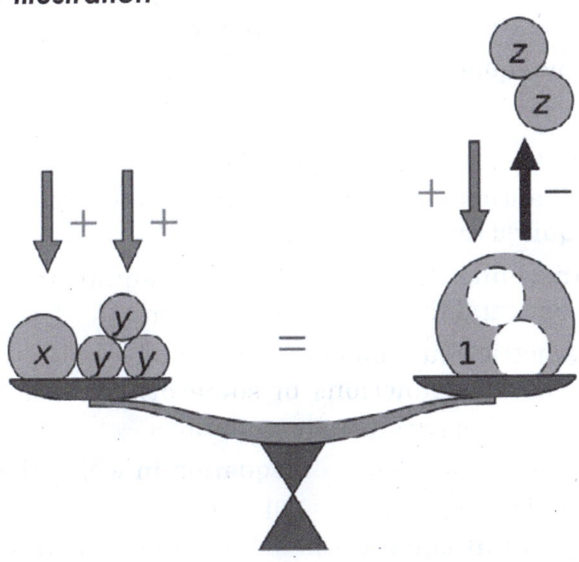

$$x + 3y = 1 - 2z$$

Figure: *Illustration of a simple equation;* x, y, z *are real numbers, analogous to weights.*

A weighing scale, balance, or seesaw is often presented as an analogy to an equation.

Each side of the balance corresponds to one side of the equation. Different quantities can be placed on each side: if the weights on the two sides are equal the scale balances, corresponding to an equality represented by an equation; if not, then the lack of balance corresponds to an inequality represented by an inequation.

In the illustration, x, y and z are all different quantities (in this case real numbers) represented as circular weights, and each of x, y, and z has a different weight. Addition corresponds to adding weight, while subtraction corresponds to removing weight from what is already there. When equality holds, the total weight on each side is the same.

Types of Equations

Equations can be classified according to the types of operations and quantities involved. Important types include:

- An algebraic equation or polynomial equation is an equation in which both sides are polynomials. These are further classified by degree:
  - o linear equation for degree one
  - o quadratic equation for degree two
  - o cubic equation for degree three
  - o quartic equation for degree four
  - o quintic equation for degree five
- A Diophantine equation is an equation where the unknowns are required to be integers
- A transcendental equation is an equation involving a transcendental function of its unknowns
- A parametric equation is an equation for which the solutions are sought as functions of some other variables, called parameters appearing in the equations
- A functional equation is an equation in which the unknowns are functions rather than simple quantities
- A differential equation is a functional equation involving derivatives of the unknown functions
- An integral equation is a functional equation involving the antiderivatives of the unknown functions
- An integro-differential equation is a functional equation involving both the derivatives and the antiderivatives of the unknown functions
- A difference equation is an equation where the unknown is a function f which occurs in the equation through $f(x)$, $f(x-1)$, ..., $f(x-k)$, for some whole integer k called the *order* of the equation. If x is restricted to be an integer, a difference equation is the same as a recurrence relation

Identities

An identity is a statement resembling an equation which is true for all possible values of the variable(s) it contains. Many identities are known, especially in trigonometry. Probably the best known example is: $\sin^2(\theta)+\cos^2(\theta)=1,$, which is true for all values of *θ*.

In the process of solving an equation, it is often useful to combine it with an identity to produce an equation which is more easily soluble. For example, to solve the equation:

$3\sin(\theta)\cos(\theta)=1,$ *where* θ *is known to be between zero and 45 degrees,*

use the identity: $\sin(2\theta)=2\sin(\theta)\cos(\theta)$, so the above equation becomes:

$$\frac{3}{2}\sin(2\theta)=1$$

Whence:

$$\theta=\frac{1}{2}\arcsin\left(\frac{2}{3}\right)\approx 20.9^{\circ}.$$

Properties

Two equations or two systems of equations are *equivalent* if they have the same set of solutions. The following operations transform an equation or a system into an equivalent one:

- Adding or subtracting the same quantity to both sides of an equation. This shows that every equation is equivalent to an equation in which the right-hand side is zero.
- Multiplying or dividing both sides of an equation by a non-zero constant.
- Applying an identity to transform one side of the equation. For example, expanding a product or factoring a sum.
- For a system: adding to both sides of an equation the corresponding side of another equation, multiplied by the same quantity.

If some function is applied to both sides of an equation, the resulting equation has the solutions of the initial equation among its solutions, but may have further solutions called extraneous solutions. For example, the equation $x=1$ has the solution $x=1$. Raising both sides to the exponent of 2 (which means applying the function $f(s)=s^2$ to both sides of the equation) changes the equation to $x^2=1$, which not only has the previous solution but also introduces the extraneous solution, $x = -1$. Moreover, If the function is not defined at some values (such as $1/x$, which is not defined for $x = 0$), solutions existing at those values may be lost. Thus, caution must be exercised when applying such a transformation to an equation.

The above transformations are the basis of most elementary methods for equation solving as well as some less elementary ones, like Gaussian elimination.

Graph of Linear Inequalities

In mathematics a linear inequality is an inequality which involves a linear function. A linear inequality contains one of the symbols of inequality:

- < is less than
- > is greater than
- ≤ is less than or equal to
- e– is greater than or equal to
- ≠ is not equal to

A linear inequality looks exactly like a linear equation, with the inequality sign replacing the equality sign.

Linear Inequalities of Real Numbers

Two-dimensional linear inequalities:

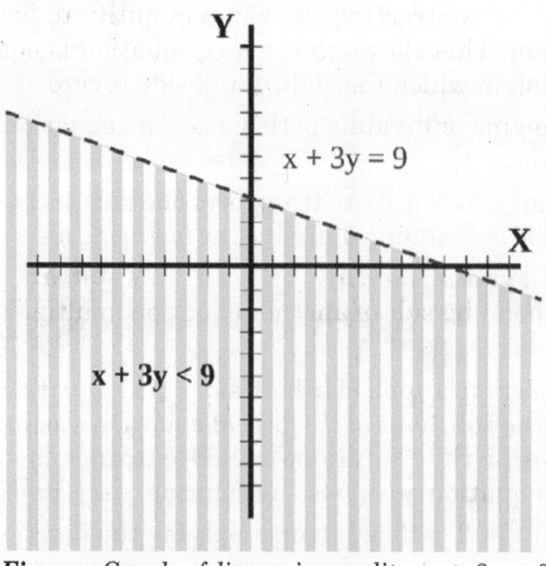

Figure: *Graph of linear inequality: $x + 3y < 9$*

Two-dimensional linear inequalities are expressions in two variables of the form:

$$ax + by < c \text{ and } ax + by \geq c,$$

where the inequalities may either be strict or not. The solution set of such an inequality can be graphically represented by a half-plane (all the points on one "side" of a fixed line) in the Euclidean plane. The line that determines the half-planes ($ax + by = c$) is not included in the solution set when the inequality is strict. A simple procedure to determine which half-plane is in the solution set is to calculate the value of $ax + by$ at a point (x_0, y_0) which is not on the line and observe whether or not the inequality is satisfied.

For example, to draw the solution set of $x + 3y < 9$, one first draws the line with equation $x + 3y = 9$ as a dotted line, to indicate that the line is not included in the solution set since the inequality is strict. Then, pick a convenient point not on the line, such as (0,0). Since $0 + 3(0) = 0 < 9$, this point is in the solution set, so the half-plane containing this point (the half-plane "below" the line) is the solution set of this linear inequality.

Linear Inequalities in General Dimensions

In R^n linear inequalities are the expressions that may be written in the form

$$f(\overline{x}) < b \text{ or } f(\overline{x}) \leq b,$$

where f is a linear form (also called a *linear functional*), $\overline{x} = (x_1, x_2, \ldots, x_n)$ and b a constant real number.

More concretely, this may be written out as

$$a_1x_1 + a_2x_2 + \cdots + a_nx_n < b,$$

or

$$a_1x_1 + a_2x_2 + \cdots + a_nx_n \leq b.$$

Here $x_1, x_2, \ldots, x_n$ are called the unknowns, and $a_1, a_2, \ldots, a_n$ are called the coefficients.

Alternatively, these may be written as

$$g(x) < 0 \text{ or } g(x) \leq 0,$$

where g is an affine function.

That is

$$a_0 + a_1x_1 + a_2x_2 + \cdots + a_nx_n < 0,$$

or

$$a_0 + a_1x_1 + a_2x_2 + \cdots + a_nx_n \leq 0.$$

Note that any inequality containing a "greater than" or a "greater than or equal" sign can be rewritten with a "less than" or "less than or equal" sign, so there is no need to define linear inequalities using those signs.

Systems of Linear Inequalities

A system of linear inequalities is a set of linear inequalities in the same variables:

$$\begin{array}{ccccccccc} a_{11}x_1 & + & a_{12}x_2 & + \cdots + & a_{1n}x_n & \leq & b_1 \\ a_{21}x_1 & + & a_{22}x_2 & + \cdots + & a_{2n}x_n & \leq & b_2 \\ \vdots & & \vdots & & \vdots & & \vdots \\ a_{m1}x_1 & + & a_{m2}x_2 & + \cdots + & a_{mn}x_n & \leq & b_m \end{array}$$

Here $x_1, x_2, ..., x_n$ are the unknowns, $a_{11}, a_{12}, ..., a_{mn}$ are the coefficients of the system, and $b_1, b_2, ..., b_m$ are the constant terms.

This can be concisely written as the matrix inequality:

$Ax \leq b$

where A is an $m \times n$ matrix, x is an $n \times 1$ column vector of variables, and b is an $m \times 1$ column vector of constants.

In the above systems both strict and non-strict inequalities may be used.

Not all systems of linear inequalities have solutions.

Applications

Polyhedra: The set of solutions of a real linear inequality constitutes a half-space of the n-dimensional real space, one of the two defined by the corresponding linear equation.

The set of solutions of a system of linear inequalities corresponds to the intersection of the half-spacess defined by individual inequalities. It is a convex set, since the half-spaces are convex sets, and the intersection of a set of convex sets is also convex. In the non-degenerate cases this convex set is a convex polyhedron (possibly unbounded, e.g., a half-space, a slab between two parallel half-spaces or a polyhedral cone). It may also be empty or a convex polyhedron of lower dimension confined to an affine subspace of the n-dimensional space R^n.

Chapter 3

Basic Concepts of Permutations and Combinations

In mathematics, the notion of permutation relates to the act of rearranging, or permuting, all the members of a set into some sequence or order (unlike combinations, which are selections of some members of the set where order is disregarded). For example, written as tuples, there are six permutations of the set {1,2,3}, namely: (1,2,3), (1,3,2), (2,1,3), (2,3,1), (3,1,2), and (3,2,1). As another example, an anagram of a word, all of whose letters are different, is a permutation of its letters. The study of permutations of finite sets is a topic in the field of combinatorics.

Figure: *Each of the six rows is a different permutation of three distinct balls*

Permutations occur, in more or less prominent ways, in almost every area of mathematics. They often arise when different orderings on certain finite sets are considered, possibly only because one wants to ignore such orderings and needs to know how many configurations are thus identified. For similar reasons permutations arise in the study of sorting algorithms in computer science.

The number of permutations of n distinct objects is n factorial usually written as $n!$, which means the product of all positive integers less than or equal to n.

Figure: *In the popular puzzle Rubik's cube invented in 1974 by Erno" Rubik, each turn of the puzzle faces creates a permutation of the surface colors.*

In algebra and particularly in group theory, a permutation of a set S is defined as a bijection from S to itself. That is, it is a function from S to S for which every element occurs exactly once as an image value. This is related to the rearrangement of the elements of S in which each element s is replaced by the corresponding $f(s)$. The collection of such permutations form a group called the symmetric group of S. The key to this group's structure is the fact that the composition of two permutations (performing two given rearrangements in succession) results in another rearrangement. Permutations may act on structured objects by rearranging their components, or by certain replacements (substitutions) of symbols.

In elementary combinatorics, the k-permutations, or partial permutations, are the ordered arrangements of k distinct elements selected from a set. When k is equal to the size of the set, these are the permutations of the set.

History

The rule to determine the number of permutations of n objects was known in Indian culture at least as early as around 1150: the Lilavati by the Indian mathematician Bhaskara II contains a passage that translates to

The product of multiplication of the arithmetical series beginning and increasing by unity and continued to the number of places, will be the variations of number with specific figures.

Fabian Stedman in 1677 described factorials when explaining the number of permutations of bells in change ringing. Starting from two bells: "first, *two* must be admitted to be varied in two ways" which he illustrates by showing 1 2 and 2 1. He then explains that with three bells there are "three times two figures to be produced out of three" which again is illustrated. His explanation involves "cast away 3, and 1.2 will remain; cast away 2, and 1.3 will remain; cast away 1, and 2.3 will remain". He then moves on to four bells and repeats the casting away argument showing that there will be four different sets of three. Effectively this is an recursive process. He continues with five bells using the "casting away" method and tabulates the resulting 120 combinations. At this point he gives up and remarks:

> *Now the nature of these methods is such, that the changes on one number comprehends the changes on all lesser numbers, ... insomuch that a compleat Peal of changes on one number seemeth to be formed by uniting of the compleat Peals on all lesser numbers into one entire body;*

Stedman widens the consideration of permutations; he goes on to consider the number of permutations of the letters of the alphabet and horses from a stable of 20.

A first case in which seemingly unrelated mathematical questions were studied with the help of permutations occurred around 1770, when Joseph Louis Lagrange, in the study of polynomial equations, observed that properties of the permutations of the roots of an equation are related to the possibilities to solve it. This line of work ultimately resulted, through the work of Évariste Galois, in Galois theory, which gives a complete description of what is possible and impossible with respect to solving polynomial equations (in one unknown) by radicals. In modern mathematics there are many similar situations in which understanding a problem requires studying certain permutations related to it.

Definition and Usage

There are two common ways of regarding permutations. They are completely equivalent and either form is readily converted to the other. Which form is preferable depends on the type of questions being asked about the permutations. Some disciplines use one form more predominantly than the other.

The first way to regard permutations of a set S (which can even be applied to infinite sets) is to define them as the bijections from S to itself. Thus, the permutations are being thought of as functions and so, can be composed with each other, forming groups of permutations. From this viewpoint, the elements of S have no special properties and are just being used as convenient names for the objects being moved around according to the bijection.

In Cauchy's *two-line notation*, one lists the elements of S in the first row, and for each one its image under the permutation below it in the second row. For instance, a particular permutation of the set {1,2,3,4,5} can be written as:

$$\sigma = \begin{pmatrix} 1 & 2 & 3 & 4 & 5 \\ 2 & 5 & 4 & 3 & 1 \end{pmatrix};$$

this means that σ satisfies $\sigma(1)=2$, $\sigma(2)=5$, $\sigma(3)=4$, $\sigma(4)=3$, and $\sigma(5)=1$. There is no special order that the elements of S appearing in the first row have to appear in. This permutation could also be written as:

$$\sigma = \begin{pmatrix} 3 & 2 & 5 & 1 & 4 \\ 4 & 5 & 1 & 2 & 3 \end{pmatrix}.$$

However, if there is a "natural" order that the elements of S can be placed in, say $x_1, x_2, \ldots, x_n$, then under the assumption that the first row of the permutation,

$$\begin{pmatrix} x_1 & x_2 & x_3 & \cdots & x_n \\ \sigma(x_1) & \sigma(x_2) & \sigma(x_3) & \cdots & \sigma(x_n) \end{pmatrix},$$

is in this natural order, the first row need not be written. Thus, under this assumption, the permutation can also be written in one-line notation as $(\sigma(x_1)\,\sigma(x_2)\,\sigma(x_3)\cdots\sigma(x_n))$, which is the second common way of representing permutations. That is, a permutation of the set S is an *ordered arrangement* (or listing, or linearly ordered arrangement, or sequence without repetition) of the elements of S. The permutation σ above would then be given by (2 5 4 3 1) since the natural order (1 2 3 4 5) would be assumed. (It is typical to use commas to separate these entries only if some have two or more digits.) This form of representation is common in elementary combinatorics, computer science and those areas of combinatorics that are closely related to it.

In many applications where the elements of S will be compared to each other, this is the preferred form of permutation representation (such applications require S to be a totally ordered set.)

There are n! permutations of a finite set S having n elements.

Circular Permutations

Permutations, when considered as arrangements, are sometimes referred to as *linearly ordered* arrangements. In these arrangements there is a first element, a second element, and so on. If, however, the objects are arranged in a circular manner this distinguished ordering no longer exists, that is, there is no "first element" in the arrangement, any element can be considered as the start of the arrangement. The arrangements of objects in a circular manner are called circular permutations.

Two circular permutations are equivalent if one can be rotated into the other (that is, cycled without changing the relative positions of the elements). The following two circular permutations on four letters are considered to be the same.

```
   1            4
 4   3        2   1
   2            3
```

The circular arrangements are to be read counterclockwise, so the following two are not equivalent since no rotation can bring one to the other.

```
   1            1
 4   3        3   4
   2            2
```

The number of circular permutations of a set S with n elements is $(n - 1)!$.

k-Permutations of n

There is also a weaker meaning of the term "permutation" that is sometimes used in elementary combinatorics texts, designating those ordered arrangements in which no element occurs more than once, but without the requirement of using all the elements from a given set. These are not permutations except in special cases, but are natural generalizations of the ordered arrangement concept.

Indeed this use often involves considering arrangements of a fixed length k of elements taken from a given set of size n, in other words, these k-permutations of n are the different ordered arrangements of a k-element subset of an n-set (sometimes called variations in the older literature.)

These objects are also known as partial permutations or as sequences without repetition, terms that avoid confusion with the other, more common, meaning of "permutation". The number of such k-permutations of n is denoted variously by such symbols as ${}_nP_k$, ${}^nP_k$, $P_{n,k}$, or $P(n,k)$, and its value is given by the product

$$P(n,k) = n\cdot(n-1)\cdot(n-2)\cdots(n-k+1)$$

which is 0 when $k > n$, and otherwise is equal to

$$\frac{n!}{(n-k)!}.$$

The product is well defined without the assumption that is a non-negative integer and is of importance outside combinatorics as well; it is known as the Pochhammer symbol $(n)_k$ or as the k-th falling factorial power n^{k} of n.

This usage of the term "permutation" is closely related to the concept of a combination of a finite set. A k-element combination of an n-set S is a k element subset of S, the elements of which are not ordered. By taking all the k element subsets of S and ordering each of them in all possible ways we obtain all the k-permutations of S. The number of k-combinations of an n-set, $C(n,k)$, is related to the number of k-permutations of n by:

$$C(n,k) = \frac{P(n,k)}{P(k,k)} = \frac{n!}{(n-k)!k!}.$$

Permutations of Multisets

If M is a finite multiset, then a multiset permutation is an ordered arrangement of elements of M in which each element appears exactly as often as is its multiplicity in M. An anagram of a word having some repeated letters is an example of a multiset permutation.

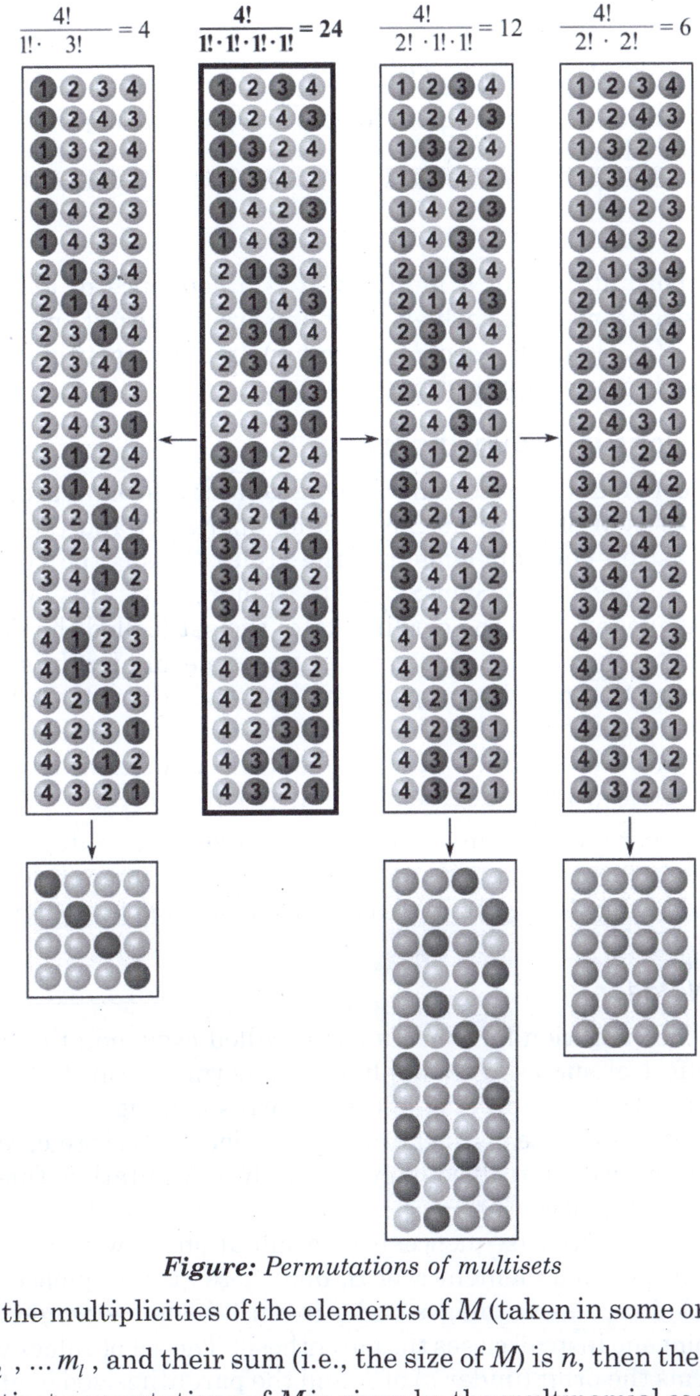

***Figure:** Permutations of multisets*

If the multiplicities of the elements of M (taken in some order) are $m_1, m_2, \ldots m_l$, and their sum (i.e., the size of M) is n, then the number of multiset permutations of M is given by the multinomial coefficient,

$$\binom{n}{m_1, m_2, \ldots, m_l} = \frac{n!}{m_1!m_2!\cdots m_l!}.$$

For example, the number of distinct anagrams of the word MISSISSIPPI is:

$$\frac{11!}{1!4!4!2!}.$$

A k-permutation of a multiset M is a sequence of length k of elements of M in which each element appears *at most* its multiplicity in M times (an element's *repetition number*). Infinite repetition numbers are allowed in some applications.

Permutations in Group Theory

The set of all permutations of any given set S forms a group, with the composition of maps as the product operation and the identity function as the neutral element of the group. This is the symmetric group of S, denoted by Sym(S). Up to isomorphism, this symmetric group only depends on the cardinality of the set (called the *degree* of the group), so the nature of elements of S is irrelevant for the structure of the group. Symmetric groups have been studied mostly in the case of finite sets, so, confined to this case, one can assume without loss of generality that $S = \{1,2,\ldots,n\}$ for some natural number n. This is then the symmetric group of degree n, usually written as S_n.

Any subgroup of a symmetric group is called a permutation group. By Cayley's theorem any group is isomorphic to some permutation group, and every finite group to a subgroup of some finite symmetric group.

Cycle Notation

Another notation for permutations called cycle notation focuses on the effect of successively applying the permutation. It expresses the permutation as a product of cycles corresponding to the orbits of the permutation; since distinct orbits are disjoint, this is referred to as "the decomposition into disjoint cycles" of the permutation. Due to the likely possibility of confusion, cycle notation is not used in conjunction with one-line notation (sequences) for permutations. It works as follows: starting from some element x of S, one writes the sequence $(x\ \sigma(x)\ \sigma(\sigma(x))\ \ldots)$ of successive images under σ, until the image would be x, at which point one instead closes the parenthesis. The set of values written down forms the orbit (under σ) of x, and the parenthesized expression gives the corresponding cycle of σ. One then continues choosing an element y of S that is not in the orbit already written down, and writes down the corresponding cycle, and so on until all elements of S belong to some cycle written down. Since for every new cycle the starting point can be chosen in different ways, there are in general many different cycle notations for the same permutation; for the example above one has for instance

$$\begin{pmatrix} 1 & 2 & 3 & 4 & 5 \\ 2 & 5 & 4 & 3 & 1 \end{pmatrix} = (1\ \ 2\ \ 5)(3\ \ 4) = (3\ \ 4)(1\ \ 2\ \ 5) = (3\ \ 4)(5\ \ 1\ \ 2).$$

Each cycle $(x_1\, x_2 \ldots x_k)$ of σ, denotes a permutation in its own right, namely the one that takes the same values as σ on this orbit (so it maps x_i to x_{i+1} for $i < k$, and x_k to x_1), while mapping all other elements of S to themselves. The size k of the orbit is called the *length* of the cycle, and a cycle of length k is called a k-cycle. Any 1-cycle is the identity permutation and so, they are all the same permutation. Distinct orbits of σ are by definition disjoint, so the corresponding cycles commute (as elements of the permutation group), and σ is the product of its cycles (taken in any order). Therefore the concatenation of cycles in the cycle notation is interpreted as denoting composition (product) of permutations, and writing a permutation as a product of its cycles is called a *decomposition into cycles* of the permutation. This decomposition is essentially unique: apart from the reordering the cycles in the product, there are no other ways to write σ as a product of cycles. The cycle notation is less unique, since each individual cycle can be written in different ways, as in the example above where (5 1 2) denotes the same cycle as (1 2 5) or (2 5 1) (though note that (5 2 1) denotes a different cycle). In writing a permutation as a product of its cycles it is typical, but not required, to suppress the writing of 1-cycles when no confusion can arise.

An orbit of size 1 (more precisely, the element of S in a 1-cycle) is called a fixed point of the permutation. A permutation that has no fixed point is called a derangement. Cycles of length two are called transpositions; such permutations merely exchange the place of two elements. Since the orbits of a permutation partition the set S, for a finite set of size n, the lengths of the cycles of a permutation σ form a partition of n called the cycle type of σ. There is a "1" in the cycle type for every fixed point of σ, a "2" for every transposition, and so on. The cycle type of $\beta = (1\ 2\ 5)(3\ 4)(6\ 8)(7)$, is (3,2,2,1) which is sometimes written in a more compact form as $(1^1, 2^2, 3^1)$.

Permutation groups have more structure than abstract groups, different realisations of a group as a permutation group need not be equivalent for this additional structure. For instance S_3 is naturally a permutation group, in which any transposition has cycle type (2,1), but the proof of Cayley's theorem realises S_3 as a subgroup of S_6 (namely the permutations of the 6 elements of S_3 itself), in which permutation group transpositions have cycle type (2,2,2). So in spite of Cayley's theorem, the study of permutation groups differs from the study of abstract groups.

Product and Inverse

The product of two permutations is defined as their composition as functions, in other words $\sigma \cdot \pi$ is the function that maps any element x of the set to $\sigma(\pi(x))$. Note that the rightmost permutation is applied to the argument first, because of the way function application is written. Some authors prefer the leftmost factor acting first, but to that end permutations must be written to the *right* of their argument, for instance as an exponent, where σ acting on x is written x^σ; then the product is defined by $x^{\sigma \cdot \pi} = (x^\sigma)^\pi$. However this gives a *different* rule for multiplying permutations; this article uses the definition where the rightmost permutation is applied first.

Since the composition of two bijections always gives another bijection, the product of two permutations is again a permutation. In two-line notation, the product of two permutations is obtained by rearranging the columns of the second (leftmost) permutation so that its first row is identical with the second row of the first (rightmost) permutation. The product can then be written as the first row of the first permutation over the second row of the modified second permutation. For example, given the permutations,

$$P = \begin{pmatrix} 1 & 2 & 3 & 4 & 5 \\ 2 & 4 & 1 & 3 & 5 \end{pmatrix} \quad \text{and} \quad Q = \begin{pmatrix} 1 & 2 & 3 & 4 & 5 \\ 5 & 4 & 3 & 2 & 1 \end{pmatrix},$$

the product QP is:

$$QP = \begin{pmatrix} 1 & 2 & 3 & 4 & 5 \\ 5 & 4 & 3 & 2 & 1 \end{pmatrix}\begin{pmatrix} 1 & 2 & 3 & 4 & 5 \\ 2 & 4 & 1 & 3 & 5 \end{pmatrix} = \begin{pmatrix} 2 & 4 & 1 & 3 & 5 \\ 4 & 2 & 5 & 3 & 1 \end{pmatrix}\begin{pmatrix} 1 & 2 & 3 & 4 & 5 \\ 2 & 4 & 1 & 3 & 5 \end{pmatrix} = \begin{pmatrix} 1 & 2 & 3 & 4 & 5 \\ 4 & 2 & 5 & 3 & 1 \end{pmatrix}.$$

In cyclic notation this same product would be given by:

$$Q \cdot P = (15)(24) \cdot (1243) = (1435).$$

Since function composition is associative, so is the product operation on permutations: $(\sigma \cdot \pi) \cdot \rho = \sigma \cdot (\pi \cdot \rho)$. Therefore, products of more than two permutations are usually written without adding parentheses to express grouping; they are also usually written without a dot or other sign to indicate multiplication. The identity permutation, which maps every element of the set to itself, is the neutral element for this product. In two-line notation, the identity is

$$\begin{pmatrix} 1 & 2 & 3 & \cdots & n \\ 1 & 2 & 3 & \cdots & n \end{pmatrix}.$$

Since bijections have inverses, so do permutations, and the inverse σ^{-1} of σ is again a permutation. Explicitly, whenever $\sigma(x)=y$ one also has $\sigma^{-1}(y)=x$. In two-line notation the inverse can be obtained by interchanging the two lines (and sorting the columns if one wishes the first line to be in a given order). For instance

$$\begin{pmatrix} 1 & 2 & 3 & 4 & 5 \\ 2 & 5 & 4 & 3 & 1 \end{pmatrix}^{-1} = \begin{pmatrix} 2 & 5 & 4 & 3 & 1 \\ 1 & 2 & 3 & 4 & 5 \end{pmatrix} = \begin{pmatrix} 1 & 2 & 3 & 4 & 5 \\ 5 & 1 & 4 & 3 & 2 \end{pmatrix}.$$

In cycle notation one can reverse the order of the elements in each cycle to obtain a cycle notation for its inverse. Thus,

$$[(125)(34)]^{-1} = (521)(43) = (152)(34).$$

Having an associative product, a neutral element, and inverses for all its elements, makes the set of all permutations of S into a group, called the symmetric group of S.

Properties

Every permutation of a finite set can be expressed as the product of transpositions. Moreover, although many such expressions for a given permutation may exist, there can never be among them both expressions with an even number and expressions with an odd number of transpositions. All permutations are then classified as even or odd, according to the parity of the transpositions in any such expression.

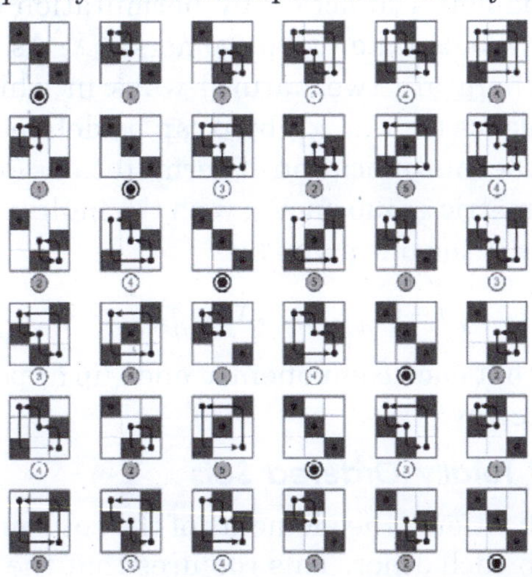

Figure: *Composition of permutations corresponds to multiplication of permutation matrices.*

Multiplying permutations written in cycle notation follows no easily described pattern, and the cycles of the product can be entirely different from those of the permutations being composed. However the cycle structure is preserved in the special case of conjugating a permutation σ by another permutation π, which means forming the product $\pi \cdot \sigma \cdot \pi^{-1}$. Here the cycle notation of the result can be obtained by taking the cycle notation for σ and applying π to all the entries in it.

Matrix Representation

One can represent a permutation of {1, 2, ..., n} as an $n \times n$ matrix. There are two natural ways to do so, but only one for which multiplications of matrices corresponds to multiplication of permutations in the same order: this is the one that associates to σ the matrix M whose entry $M_{i,j}$ is 1 if $i = \sigma(j)$, and 0 otherwise. The resulting matrix has exactly one entry 1 in each column and in each row, and is called a permutation matrix.

Here (file) is a list of these matrices for permutations of 4 elements. The Cayley table on the right shows these matrices for permutations of 3 elements.

Permutation of Components of a Sequence

As with any group, one can consider actions of a symmetric group on a set, and there are many ways in which such an action can be defined. For the symmetric group of {1, 2, ..., n} there is one particularly natural action, namely the action by permutation on the set X^n of sequences of n symbols taken from some set X. As with the matrix representation, there are two natural ways in which the result of permuting a sequence $(x_1, x_2, \ldots, x_n)$ by σ can be defined, but only one is compatible with the multiplication of permutations (so as to give a left action of the symmetric group on X^n); with the multiplication rule used in this article this is the one given by

$$\sigma \cdot (x_1, \ldots, x_n) = (x_{\sigma^{-1}(1)}, \ldots, x_{\sigma^{-1}(n)}).$$

This means that each component x_i ends up at position $\sigma(i)$ in the sequence permuted by σ.

Permutations of Totally Ordered Sets

In some applications, the elements of the set being permuted will be compared with each other. This requires that the set S has a total order so that any two elements can be compared. The set {1, 2, ..., n} is totally ordered by the usual "≤" relation and so it is the most frequently used set in these applications, but in general, any totally ordered set will do. In these applications, the ordered arrangement view of a permutation is needed to talk about the *positions* in a permutation.

Here are a number of properties that are directly related to the total ordering of S.

Ascents, Descents and Runs

An *ascent* of a permutation σ of n is any position $i < n$ where the following value is bigger than the current one. That is, if $\sigma = \sigma_1\sigma_2...\sigma_n$, then i is an ascent if $\sigma_i < \sigma_{i+1}$.

For example, the permutation 3452167 has ascents (at positions) 1,2,5,6.

Similarly, a *descent* is a position $i < n$ with $\sigma_i > \sigma_{i+1}$, so every i with $1 \le i < n$ either is an ascent or is a descent of σ.

The number of permutations of n with k ascents is the Eulerian number $\left\langle {n \atop k} \right\rangle$; this is also the number of permutations of n with k descents.

An *ascending run* of a permutation is a nonempty increasing contiguous subsequence of the permutation that cannot be extended at either end; it corresponds to a maximal sequence of successive ascents (the latter may be empty: between two successive descents there is still an ascending run of length 1). By contrast an *increasing subsequence* of a permutation is not necessarily contiguous: it is an increasing sequence of elements obtained from the permutation by omitting the values at some positions. For example, the permutation 2453167 has the ascending runs 245, 3, and 167, while it has an increasing subsequence 2367.

If a permutation has $k - 1$ descents, then it must be the union of k ascending runs. Hence, the number of permutations of n with k ascending runs is the same as the number $\left\langle {n \atop k-1} \right\rangle$ of permutations with $k - 1$ descents.

Inversions

An inversion of a permutation σ is a pair (i,j) of positions where the entries of a permutation are in the opposite order: $i < j$ and $\sigma_i > \sigma_j$. So a descent is just an inversion at two adjacent positions. For example, the permutation $\sigma = 23154$ has three inversions: (1,3), (2,3), (4,5), for the pairs of entries (2,1), (3,1), (5,4).

Figure: *In the 15 puzzle the goal is to get the squares in ascending order. Initial positions which have an odd number of inversions are impossible to solve.*

Sometimes an inversion is defined as the pair of values (σ_i, σ_j) itself whose order is reversed; this makes no difference for the *number* of inversions, and this pair (reversed) is also an inversion in the above sense for the inverse permutation σ^{-1}. The number of inversions is an important measure for the degree to which the entries of a permutation are out of order; it is the same for σ and for σ^{-1}. To bring a permutation with k inversions into order (i.e., transform it into the identity permutation), by successively applying (right-multiplication by) adjacent transpositions, is always possible and requires a sequence of k such operations. Moreover any reasonable choice for the adjacent transpositions will work: it suffices to choose at each step a transposition of i and $i + 1$ where i is a descent of the permutation as modified so far (so that the transposition will remove this particular descent, although it might create other descents). This is so because applying such a transposition reduces the number of inversions by 1; also note that as long as this number is not zero, the permutation is not the identity, so it has at least one descent. Bubble sort and insertion sort can be interpreted as particular instances of this procedure to put a sequence into order. Incidentally this procedure proves that any permutation σ can be written as a product of adjacent transpositions; for this one may simply reverse any sequence of such transpositions that transforms σ into the identity. In fact, by enumerating all sequences of adjacent transpositions that would transform σ into the identity, one obtains (after reversal) a *complete* list of all expressions of minimal length writing σ as a product of adjacent transpositions.

The number of permutations of n with k inversions is expressed by a Mahonian number, it is the coefficient of X^k in the expansion of the product

$$\prod_{m=1}^{n}\sum_{i=0}^{m-1} X^i = 1(1+X)(1+X+X^2)\cdots(1+X+X^2+\cdots+X^{n-1}),$$

which is also known (with q substituted for X) as the q-factorial $[n]_q!$. The expansion of the product appears in Necklace (combinatorics).

Permutations in Computing

Numbering Permutations

One way to represent permutations of n is by an integer N with $0 \le N < n!$, provided convenient methods are given to convert between the number and the representation of a permutation as an ordered arrangement (sequence). This gives the most compact representation of arbitrary permutations, and in computing is particularly attractive when n is small enough that N can be held in a machine word; for 32-bit words this means $n \le 12$, and for 64-bit words this means $n \le 20$. The conversion can be done via the intermediate form of a sequence of numbers $d_n, d_{n-1}, \ldots, d_2, d_1$, where d_i is a non-negative integer less than i (one may omit d_1, as it is always 0, but its presence makes the subsequent conversion to a permutation easier to describe). The first step then is simply expression of N in the factorial number system, which is just a particular mixed radix representation, where for numbers up to $n!$ the bases for successive digits are $n, n-1, \ldots, 2, 1$. The second step interprets this sequence as a Lehmer code or (almost equivalently) as an inversion table.

Rothe diagram for $\sigma = (6,3,8,1,4,9,7,2,5)$

$i \backslash \sigma_i$	1	2	3	4	5	6	7	8	9	Lehmer code
1	×	×	×	×	×	•				$d_9 = 5$
2	×	×	•							$d_8 = 2$
3	×	×		×	×		×	•		$d_7 = 5$
4	•									$d_6 = 0$
5		×		•						$d_5 = 1$
6		×			×		×		•	$d_4 = 3$
7		×			×		•			$d_3 = 2$
8		•								$d_2 = 0$
9					•					$d_1 = 0$
inversion table	3	6	1	2	4	0	2	0	0	

In the Lehmer code for a permutation σ, the number d_n represents the choice made for the first term σ_1, the number d_{n-1} represents the choice made for the second term σ_2 among the remaining $n-1$ elements of the set, and so forth. More precisely, each d_{n+1-i} gives the number of *remaining* elements strictly less than the term σ_i. Since those remaining elements are bound to turn up as some later term σ_j, the digit d_{n+1-i} counts the *inversions* (i,j) involving i as smaller index (the number of values j for which $i < j$ and $\sigma_i > \sigma_j$). The inversion table for σ is quite similar, but here d_{n+1-k} counts the number of inversions (i,j) where $k = \sigma_j$ occurs as the smaller of the two values appearing in inverted order. Both encodings can be visualized by an n by n Rothe diagram (named after Heinrich August Rothe) in which dots at (i,σ_i) mark the entries of the permutation, and a cross at (i,σ_j) marks the inversion (i,j); by the definition of inversions a cross appears in any square that comes both before the dot (j,σ_j) in its column, and before the dot (i,σ_i) in its row. The Lehmer code lists the numbers of crosses in successive rows, while the inversion table lists the numbers of crosses in successive columns; it is just the Lehmer code for the inverse permutation, and vice versa.

To effectively convert a Lehmer code $d_n, d_{n-1}, \ldots, d_2, d_1$ into a permutation of an ordered set S, one can start with a list of the elements of S in increasing order, and for i increasing from 1 to n set σ_i to the element in the list that is preceded by d_{n+1-i} other ones, and remove that element from the list. To convert an inversion table $d_n, d_{n-1}, \ldots, d_2, d_1$ into the corresponding permutation, one can traverse the numbers from d_1 to d_n while inserting the elements of S from largest to smallest into an initially empty sequence; at the step using the number d from the inversion table, the element from S inserted into the sequence at the point where it is preceded by d elements already present. Alternatively one could process the numbers from the inversion table and the elements of S both in the opposite order, starting with a row of n empty slots, and at each step place the element from S into the empty slot that is preceded by d other empty slots.

Converting successive natural numbers to the factorial number system produces those sequences in lexicographic order (as is the case with any mixed radix number system), and further converting them to permutations preserves the lexicographic ordering, provided the Lehmer code interpretation is used (using inversion tables, one gets a different ordering, where one starts by comparing permutations by the *place* of their entries 1 rather than by the value of their first entries). The sum of the numbers in the factorial number system representation gives the number of inversions of the permutation, and the parity of that sum gives the signature of the permutation. Moreover the positions of the zeroes in the inversion table give the values of left-to-right maxima of the permutation (in the example 6, 8, 9) while the positions of the zeroes in the Lehmer code are the positions of the right-to-left minima (in the example positions the 4, 8, 9 of the values 1, 2, 5); this allows computing the distribution of such extrema among all permutations. A permutation with Lehmer code $d_n, d_{n-1}, \ldots, d_2, d_1$ has an ascent $n - i$ if and only if d_i e– d_{i+1}.

Algorithms to Generate Permutations

In computing it may be required to generate permutations of a given sequence of values. The methods best adapted to do this depend on whether one wants some randomly chosen permutations, or all permutations, and in the latter case if a specific ordering is required. Another question is whether possible equality among entries in the given sequence is to be taken into account; if so, one should only generate distinct multiset permutations of the sequence.

An obvious way to generate permutations of *n* is to generate values for the Lehmer code (possibly using the factorial number system representation of integers up to *n*!), and convert those into the corresponding permutations. However, the latter step, while straightforward, is hard to implement efficiently, because it requires *n* operations each of selection from a sequence and deletion from it, at an arbitrary position; of the obvious representations of the sequence as an array or a linked list, both require (for different reasons) about $n^2/4$ operations to perform the conversion. With *n* likely to be rather small (especially if generation of all permutations is needed) that is not too much of a problem, but it turns out that both for random and for systematic generation there are simple alternatives that do considerably better. For this reason it does not seem useful, although certainly possible, to employ a special data structure that would allow performing the conversion from Lehmer code to permutation in O(n log n) time.

Random Generation of Permutations

For generating random permutations of a given sequence of *n* values, it makes no difference whether one means apply a randomly selected permutation of *n* to the sequence, or choose a random element from the set of distinct (multiset) permutations of the sequence. This is because, even though in case of repeated values there can be many distinct permutations of *n* that result in the same permuted sequence, the number of such permutations is the same for each possible result.

Unlike for systematic generation, which becomes unfeasible for large n due to the growth of the number $n!$, there is no reason to assume that n will be small for random generation.

The basic idea to generate a random permutation is to generate at random one of the $n!$ sequences of integers $d_1, d_2, \ldots, d_n$ satisfying $0 \leq d_i < i$ (since d_1 is always zero it may be omitted) and to convert it to a permutation through a bijective correspondence. For the latter correspondence one could interpret the (reverse) sequence as a Lehmer code, and this gives a generation method first published in 1938 by Ronald A. Fisher and Frank Yates. While at the time computer implementation was not an issue, this method suffers from the difficulty sketched above to convert from Lehmer code to permutation efficiently. This can be remedied by using a different bijective correspondence: after using d_i to select an element among i remaining elements of the sequence (for decreasing values of i), rather than removing the element and compacting the sequence by shifting down further elements one place, one swaps the element with the final remaining element. Thus the elements remaining for selection form a consecutive range at each point in time, even though they may not occur in the same order as they did in the original sequence. The mapping from sequence of integers to permutations is somewhat complicated, but it can be seen to produce each permutation in exactly one way, by an immediate induction. When the selected element happens to be the final remaining element, the swap operation can be omitted. This does not occur sufficiently often to warrant testing for the condition, but the final element must be included among the candidates of the selection, to guarantee that all permutations can be generated.

The resulting algorithm for generating a random permutation of $a[0], a[1], \ldots, a[n-1]$ can be described as follows in pseudocode:

for i from n downto 2

do $d_i \leftarrow$ random element of $\{0, \ldots, i-1\}$

swap $a[d_i]$ and $a[i-1]$

This can be combined with the initialization of the array $a[i] = i$ as follows:

for i from 0 to $n-1$

do $d_{i+1} \leftarrow$ random element of $\{0, \ldots, i\}$

$a[i] \leftarrow a[d_{i+1}]$

$a[d_{i+1}] \leftarrow i$

If $d_{i+1} = i$, the first assignment will copy an uninitialized value, but the second will overwrite it with the correct value i.

Generation in Lexicographic Order

There are many ways to systematically generate all permutations of a given sequence. One classical algorithm, which is both simple and flexible, is based on finding the next permutation in lexicographic ordering, if it exists. It can handle repeated values, for which case it generates the distinct multiset permutations each once. Even for ordinary permutations it is significantly more efficient than generating values for the Lehmer code in lexicographic order (possibly using the factorial number system) and converting those to permutations. To use it, one starts by sorting the sequence in (weakly) increasing order (which gives its lexicographically minimal permutation), and then repeats advancing to the next permutation as long as one is found. The method goes back to Narayana Pandita in 14th century India, and has been frequently rediscovered ever since.

The following algorithm generates the next permutation lexicographically after a given permutation. It changes the given permutation in-place.

1. Find the largest index k such that $a[k] < a[k + 1]$. If no such index exists, the permutation is the last permutation.
2. Find the largest index l greater than k such that $a[k] < a[l]$.
3. Swap the value of $a[k]$ with that of $a[l]$.
4. Reverse the sequence from $a[k + 1]$ up to and including the final element $a[n]$.

For example, given the sequence [1, 2, 3, 4] which starts in a weakly increasing order, and given that the index is zero-based, the steps are as follows:

1. Index $k = 2$, because 3 is placed at an index that satisfies condition of being the largest index that is still less than $a[k + 1]$ which is 4.
2. Index $l = 3$, because 4 is the only value in the sequence that is greater than 3 in order to satisfy the condition $a[k] < a[l]$.
3. The values of $a[2]$ and $a[3]$ are swapped to form the new sequence [1,2,4,3].
4. The sequence after k-index $a[2]$ to the final element is reversed. Because only one value lies after this index (the 3), the sequence remains unchanged in this instance. Thus the lexicographic successor of the initial state is permuted: [1,2,4,3].

Following this algorithm, the next lexicographic permutation will be [1,3,2,4], and the 24th permutation will be [4,3,2,1] at which point $a[k] < a[k + 1]$ does not exist, indicating that this is the last permutation.

Generation with Minimal Changes

An alternative to the above algorithm, the Steinhaus–Johnson–Trotter algorithm, generates an ordering on all the permutations of a given sequence with the property that any two consecutive permutations in its output differ by swapping two adjacent values. This ordering on the permutations was known to 17th-century English bell ringers, among whom it was known as "plain changes".

One advantage of this method is that the small amount of change from one permutation to the next allows the method to be implemented in constant time per permutation.

The same can also easily generate the subset of even permutations, again in constant time per permutation, by skipping every other output permutation. An old paper proposed a very efficient algorithm called Heap's permutation generation algorithm. An article authored by Robert Sedgewick said that this algorithm is the fastest algorithm of generating permutations in applications.

Meandric Permutations

Meandric systems give rise to meandric permutations, a special subset of *alternate permutations*. An alternate permutation of the set $\{1,2,...,2n\}$ is a cyclic permutation (with no fixed points) such that the digits in the cyclic notation form alternate between odd and even integers. Meandric permutations are useful in the analysis of RNA secondary structure. Not all alternate permutations are meandric. A modification of Heap's algorithm has been used to generate all alternate permutations of order n (that is, of length $2n$) without generating all $(2n)!$ permutations. Generation of these alternate permutations is needed before they are analyzed to determine if they are meandric or not.

The algorithm is recursive. The following table exhibits a step in the procedure. In the previous step, all alternate permutations of length 5 have been generated. Three copies of each of these have a "6" added to the right end, and then a different transposition involving this last entry and a previous entry in an even position is applied (including the identity, i.e., no transposition).

Previous Sets	*Transposition of Digits*	*Alternate Permutations*
1-2-3-4-5-6		1-2-3-4-5-6
1-2-3-4-5-6	4,6	1-2-3-6-5-4
1-2-3-4-5-6	2,6	1-6-3-4-5-2
1-2-5-4-3-6		1-2-5-4-3-6
1-2-5-4-3-6	4,6	1-2-5-6-3-4
1-2-5-4-3-6	2,6	1-6-5-4-3-2
1-4-3-2-5-6		1-4-3-2-5-6
1-4-3-2-5-6	2,6	1-4-3-6-5-2
1-4-3-2-5-6	4,6	1-6-3-2-5-4
1-4-5-2-3-6		1-4-5-2-3-6
1-4-5-2-3-6	2,6	1-4-5-6-3-2
1-4-5-2-3-6	4,6	1-6-5-2-3-4

Software Implementations

Calculator Functions

Many scientific calculators and computing software have a built-in function for calculating the number of k-permutations of n.

- Casio and TI calculators: nPr
- HP calculators: PERM
- Mathematica: FactorialPower

Spreadsheet Functions

Most spreadsheet software also provides a built-in function for calculating the number of k-permutations of n, called PERMUT in many popular spreadsheets.

Applications

Permutations are used in the interleaver component of the error detection and correction algorithms, such as turbo codes, for example 3GPP Long Term Evolution mobile telecommunication standard uses these ideas. Such applications raise the question of fast generation of permutations satisfying certain desirable properties. One of the methods is based on the permutation polynomials. Also as a base for optimal hashing in Unique Permutation Hashing.

Combinations

In mathematics, a combination is a way of selecting members from a grouping, such that (unlike permutations) the order of selection does not matter. In smaller cases it is possible to count the number of combinations. For example given three fruits, say an apple, an orange and a pear, there are three combinations of two that can be drawn from this set: an apple and a pear; an apple and an orange; or a pear and an orange. More formally, a k-combination of a set S is a subset of k distinct elements of S. If the set has n elements, the number of k-combinations is equal to the binomial coefficient

$$\binom{n}{k} = \frac{n(n-1)\ldots(n-k+1)}{k(k-1)\ldots1},$$

which can be written using factorials as $\frac{n!}{k!(n-k)!}$ whenever $k \leq n$, and which is zero when $k > n$. The set of all k-combinations of a set S is sometimes denoted by $\binom{S}{k}$.

Combinations refer to the combination of n things taken k at a time without repetition. To refer to combinations in which repetition is allowed, the terms k-selection, k-multiset, or k-combination with repetition are often used. If, in the above example, it was possible to have two of any one kind of fruit there would be 3 more 2-selections: one with two apples, one with two oranges, and one with two pears.

Although the set of three fruits was small enough to write a complete list of combinations, with large sets this becomes impractical. For example, a poker hand can be described as a 5-combination (k = 5) of cards from a 52 card deck (n = 52). The 5 cards of the hand are all distinct, and the order of cards in the hand does not matter. There are 2,598,960 such combinations, and the chance of drawing any one hand at random is 1 / 2,598,960.

Number of k-Combinations

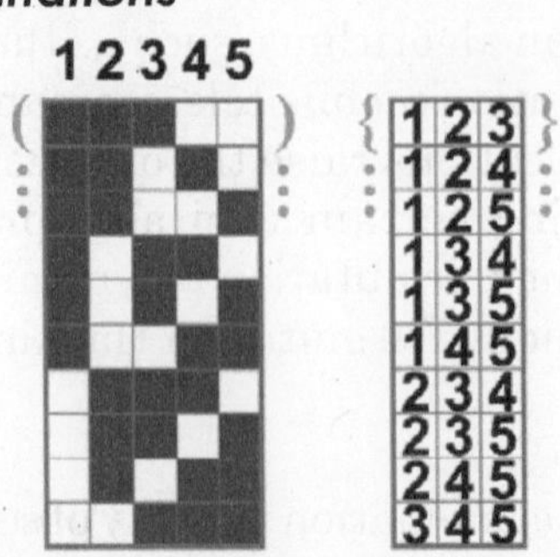

Figure: *3-element subsets of a 5-element set*

The number of k-combinations from a given set S of n elements is often denoted in elementary combinatorics texts by $C(n, k)$, or by a variation such as C_k^n, ${}_nC_k$, ${}^nC_k$ or even C_n^k (the latter form was standard in French, Russian, Chinese and Polish texts). The same number however occurs in many other mathematical contexts, where it is denoted by $\binom{n}{k}$ (often read as "n choose k"); notably it occurs as a coefficient in the binomial formula, hence its name binomial coefficient. One can define $\binom{n}{k}$ for all natural numbers k at once by the relation

$$(1+X)^n = \sum_{k\geq 0}\binom{n}{k}X^k,$$

from which it is clear that $\binom{n}{0}=\binom{n}{n}=1$ and $\binom{n}{k}=0$ for $k > n$. To see that these coefficients count k-combinations from S, one can first consider a collection of n distinct variables X_s labelled by the elements s of S, and expand the product over all elements of S:

$$\prod_{s\in S}(1+X_s);$$

it has 2^n distinct terms corresponding to all the subsets of S, each subset giving the product of the corresponding variables X_s. Now setting all of the X_s equal to the unlabelled variable X, so that the product becomes $(1 + X)^n$, the term for each k-combination from S becomes X^k, so that the coefficient of that power in the result equals the number of such k-combinations.

Binomial coefficients can be computed explicitly in various ways. To get all of them for the expansions up to $(1 + X)^n$, one can use (in addition to the basic cases already given) the recursion relation

$$\binom{n}{k}=\binom{n-1}{k-1}+\binom{n-1}{k}, \text{ for } 0<k<n,$$

which follows from $(1 + X)^n = (1 + X)^{n-1}(1 + X)$; this leads to the construction of Pascal's triangle.

For determining an individual binomial coefficient, it is more practical to use the formula

$$\binom{n}{k}=\frac{n(n-1)(n-2)\cdots(n-k+1)}{k!}.$$

The numerator gives the number of k-permutations of n, i.e., of sequences of k distinct elements of S, while the denominator gives the number of such k-permutations that give the same k-combination when the order is ignored. When k exceeds $n/2$, the above formula contains factors common to the numerator and the denominator, and canceling them out gives the relation

$$\binom{n}{k}=\binom{n}{n-k}, \text{ for } 0 \le k \le n.$$

This expresses a symmetry that is evident from the binomial formula, and can also be understood in terms of k-combinations by taking the complement of such a combination, which is an $(n-k)$-combination. Finally there is a formula which exhibits this symmetry directly, and has the merit of being easy to remember:

$$\binom{n}{k}=\frac{n!}{k!(n-k)!},$$

where $n!$ denotes the factorial of n. It is obtained from the previous formula by multiplying denominator and numerator by $(n-k)!$, so it is certainly inferior as a method of computation to that formula.

The last formula can be understood directly, by considering the $n!$ permutations of all the elements of S. Each such permutation gives a k-combination by selecting its first k elements. There are many duplicate selections: any combined permutation of the first k elements among each other, and of the final $(n-k)$ elements among each other produces the same combination; this explains the division in the formula.

From the above formulas follow relations between adjacent numbers in Pascal's triangle in all three directions:

$$\binom{n}{k}=\binom{n}{k-1}\frac{n-k+1}{k}, \text{ for } k>0,$$

$$\binom{n}{k}=\binom{n-1}{k}\frac{n}{n-k}, \text{ for } k<n,$$

$$\binom{n}{k}=\binom{n-1}{k-1}\frac{n}{k}, \text{ for } n,k>0.$$

Together with the basic cases $\binom{n}{0}=1=\binom{n}{n}$, these allow successive computation of respectively all numbers of combinations from the same set (a row in Pascal's triangle), of k-combinations of sets of growing sizes, and of combinations with a complement of fixed size $n-k$.

Example of Counting Combinations

As a concrete example, one can compute the number of five-card hands possible from a standard fifty-two card deck as:

$$\binom{52}{5} = \frac{52\times51\times50\times49\times48}{5\times4\times3\times2\times1} = \frac{311,875,200}{120} = 2,598,960.$$

Alternatively one may use the formula in terms of factorials and cancel the factors in the numerator against parts of the factors in the denominator, after which only multiplication of the remaining factors is required:

$$\begin{aligned}\binom{52}{5} &= \frac{52!}{5!47!} \\ &= \frac{52\times51\times50\times49\times48\times47!}{5\times4\times3\times2\times1\times47!} \\ &= \frac{52\times51\times50\times49\times48}{5\times4\times3\times2} \\ &= \frac{(26\times2)\times(17\times3)\times(10\times5)\times49\times(12\times4)}{5\times4\times3\times2} \\ &= 26\times17\times10\times49\times12 \\ &= 2,598,960.\end{aligned}$$

Another alternative computation, equivalent to the first, is based on writing

$$\binom{n}{k} = \frac{(n-0)}{1}\times\frac{(n-1)}{2}\times\frac{(n-2)}{3}\times\cdots\times\frac{(n-(k-1))}{k},$$

which gives

$$\binom{52}{5} = \frac{52}{1}\times\frac{51}{2}\times\frac{50}{3}\times\frac{49}{4}\times\frac{48}{5} = 2,598,960.$$

When evaluated in the following order, $52 \div 1 \times 51 \div 2 \times 50 \div 3 \times 49 \div 4 \times 48 \div 5$, this can be computed using only integer arithmetic. The reason is that when each division occurs, the intermediate result that is produced is itself a binomial coefficient, so no remainders ever occur.

Using the symmetric formula in terms of factorials without performing simplifications gives a rather extensive calculation:

Enumerating k-Combinations

One can enumerate all k-combinations of a given set S of n elements in some fixed order, which establishes a bijection from an interval of

$\binom{n}{k}$ integers with the set of those k-combinations. Assuming S is itself ordered, for instance $S = \{1,2, ...,n\}$, there are two natural possibilities for ordering its k-combinations: by comparing their smallest elements first (as in the illustrations above) or by comparing their largest elements first. The latter option has the advantage that adding a new largest element to S will not change the initial part of the enumeration, but just add the new k-combinations of the larger set after the previous ones. Repeating this process, the enumeration can be extended indefinitely with k-combinations of ever larger sets. If moreover the intervals of the integers are taken to start at 0, then the k-combination at a given place i in the enumeration can be computed easily from i, and the bijection so obtained is known as the combinatorial number system. It is also known as "rank"/"ranking" and "unranking" in computational mathematics.

There are many ways to enumerate k combinations. One way is to visit all the binary numbers less than 2^n. Chose those numbers having k nonzero bits. The positions of these 1 bits in such a number is a specific k-combination of the set $\{1,...,n\}$.

Number of Combinations with Repetition

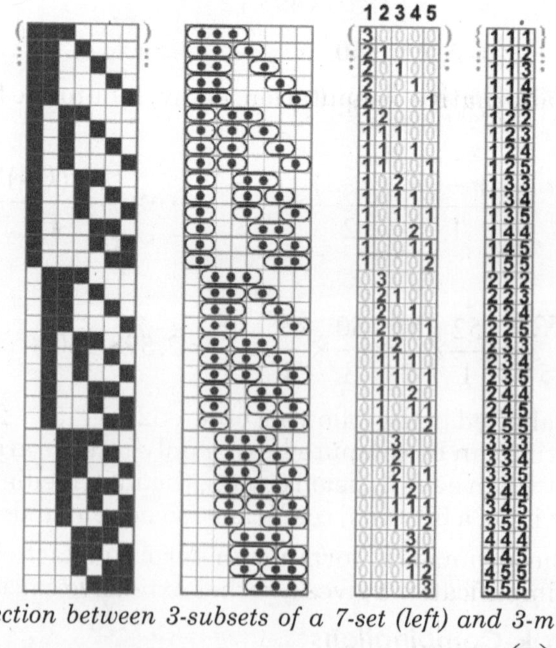

Figure: *Bijection between 3-subsets of a 7-set (left) and 3-multisets with elements from a 5-set (right). So this illustrates that* $\binom{7}{3} = \left(\!\binom{5}{3}\!\right)$.

A k-combination with repetitions, or k-multicombination, or multisubset of size k from a set S is given by a sequence of k not necessarily distinct elements of S, where order is not taken into account: two sequences of which one can be obtained from the other by permuting the terms define the same multiset. In other words, the number of ways to sample k elements from a set of n elements allowing for duplicates (i.e., with replacement) but disregarding different orderings (e.g. {2,1,2} = {1,2,2}). Associate an index to each element of S and think of the elements of S as *types* of objects, then we can let x_i denote the number of elements of type i in a multisubset. The number of multisubsets of size k is then the number of nonnegative integer solutions of the Diophantine equation:

$$x_1 + x_2 + \ldots + x_n = k.$$

If S has n elements, the number of such k-multisubsets is denoted by,

$$\left(\!\!\binom{n}{k}\!\!\right),$$

a notation that is analogous to the binomial coefficient which counts k-subsets. This expression, n multichoose k, is also given by a binomial coefficient:

$$\left(\!\!\binom{n}{k}\!\!\right) = \binom{n+k-1}{k}.$$

This relationship can be easily seen using a representation known as stars and bars. A solution of the above Diophantine equation can be represented by x_2 *stars*, a separator (a *bar*), then x_1 more stars, another separator, and so on. The total number of stars in this representation is k and the number of bars is n - 1 (since no separator is needed at the very end). Thus, a string of $k + n$ - 1 symbols (stars and bars) corresponds to a solution if there are k stars in the string. Any solution can be represented by choosing k out of $k + n$ - 1 positions to place stars and filling the remaining positions with bars. For example, the solution $x_1 = 3, x_2 = 2, x_3 = 0, x_4 = 5$ of the equation $x_1 + x_2 + x_3 + x_4 = 10$ can be represented by

★★★|★★||★★★★★

The number of such strings is the number of ways to place 10 stars in 13 positions, $\binom{13}{10} = \binom{13}{3} = 286$, which is the number of 10-multisubsets of a set with 4 elements.

As with binomial coefficients, there are several relationships between these multichoose expressions. For example, for $n \geq 1, k \geq 0$,

$$\left(\!\!\binom{n}{k}\!\!\right) = \left(\!\!\binom{k+1}{n-1}\!\!\right).$$

This identity follows from interchanging the stars and bars in the above representation.

Example of Counting Multisubsets

For example, if you have four types of donuts (n = 4) on a menu to choose from and you want three donuts (k = 3), the number of ways to choose the donuts with repetition can be calculated as

$$\left(\!\!\binom{4}{3}\!\!\right) = \binom{4+3-1}{3} = \binom{6}{3} = \frac{6\times5\times4}{3\times2\times1} = 20.$$

This result can be verified by listing all the 3-multisubsets of the set $S = \{1,2,3,4\}$. This is displayed in the following table. The second column shows the nonnegative integer solutions $[x_1, x_2, x_3, x_4]$ of the equation $x_1 + x_2 + x_3 + x_4 = 3$ and the last column gives the stars and bars representation of the solutions.

<table>
<tr><th>No.</th><th>3-Multiset</th><th>Eq. Solution</th><th>Stars and Bars</th></tr>
<tr><td>1</td><td>{1,1,1}</td><td>[3,0,0,0]</td><td>★★★|||</td></tr>
<tr><td>2</td><td>{1,1,2}</td><td>[2,1,0,0]</td><td>★★|★||</td></tr>
<tr><td>3</td><td>{1,1,3}</td><td>[2,0,1,0]</td><td>★★||★|</td></tr>
<tr><td>4</td><td>{1,1,4}</td><td>[2,0,0,1]</td><td>★★|||★</td></tr>
<tr><td>5</td><td>{1,2,2}</td><td>[1,2,0,0]</td><td>★|★★||</td></tr>
<tr><td>6</td><td>{1,2,3}</td><td>[1,1,1,0]</td><td>★|★|★|</td></tr>
<tr><td>7</td><td>{1,2,4}</td><td>[1,1,0,1]</td><td>★|★||★</td></tr>
<tr><td>8</td><td>{1,3,3}</td><td>[1,0,2,0]</td><td>★||★★|</td></tr>
<tr><td>9</td><td>{1,3,4}</td><td>[1,0,1,1]</td><td>★||★|★</td></tr>
<tr><td>10</td><td>{1,4,4}</td><td>[1,0,0,2]</td><td>★|||★★</td></tr>
<tr><td>11</td><td>{2,2,2}</td><td>[0,3,0,0]</td><td>|★★★||</td></tr>
<tr><td>12</td><td>{2,2,3}</td><td>[0,2,1,0]</td><td>|★★|★|</td></tr>
<tr><td>13</td><td>{2,2,4}</td><td>[0,2,0,1]</td><td>|★★||★</td></tr>
<tr><td>14</td><td>{2,3,3}</td><td>[0,1,2,0]</td><td>|★|★★|</td></tr>
<tr><td>15</td><td>{2,3,4}</td><td>[0,1,1,1]</td><td>|★|★|★</td></tr>
<tr><td>16</td><td>{2,4,4}</td><td>[0,1,0,2]</td><td>|★||★★</td></tr>
<tr><td>17</td><td>{3,3,3}</td><td>[0,0,3,0]</td><td>||★★★|</td></tr>
<tr><td>18</td><td>{3,3,4}</td><td>[0,0,2,1]</td><td>||★★|★</td></tr>
<tr><td>19</td><td>{3,4,4}</td><td>[0,0,1,2]</td><td>||★|★★</td></tr>
<tr><td>20</td><td>{4,4,4}</td><td>[0,0,0,3]</td><td>|||★★★</td></tr>
</table>

Number of k-Combinations for all k

The number of k-combinations for all k is the number of subsets of a set of n elements. There are several ways to see that this number is 2^n. In terms of combinations, $\sum_{0 \le k \le n} \binom{n}{k} = 2^n$, which is the sum of the nth row (counting from 0) of the binomial coefficients in Pascal's triangle. These combinations (subsets) are enumerated by the 1 digits of the set of base 2 numbers counting from 0 to $2^n - 1$, where each digit position is an item from the set of n.

Given 3 cards numbered 1 to 3, there are 8 distinct combinations (subsets), including the empty set:

$$|\{\{\};\{1\};\{2\};\{3\};\{1,2\};\{1,3\};\{2,3\};\{1,2,3\}\}| = 2^3 = 8$$

Probability: Sampling a Random Combination

There are various algorithms to pick out a random combination from a given set or list. Rejection sampling is extremely slow for large sample sizes. One way to select a k-combination efficiently from a population of size n is to iterate across each element of the population, and at each step pick that element with a dynamically changing probability of $\frac{k - \#\text{samples chosen}}{n - \#\text{samples visited}}$.

Binomial Coefficient

In mathematics, binomial coefficients are a family of positive integers that occur as coefficients in the binomial theorem. They are indexed by two nonnegative integers; the binomial coefficient indexed by n and k is usually written $\binom{n}{k}$. It is the coefficient of the x^k term in the polynomial expansion of the binomial power $(1 + x)^n$. Under suitable circumstances the value of the coefficient is given by the expression $\frac{n!}{k!(n-k)!}$. Arranging binomial coefficients into rows for successive values of n, and in which k ranges from 0 to n, gives a triangular array called Pascal's triangle.

```
          1
        1   1
      1   2   1
    1   3   3   1
  1   4   6   4   1
1   5  10  10   5   1
```

Figure: *The binomial coefficients can be arranged to form Pascal's triangle.*

This family of numbers also arises in many areas of mathematics other than algebra, notably in combinatorics. For any set containing n elements, the number of distinct k-element subsets of it that can be formed (the k-combinations of its elements) is given by the binomial coefficient $\binom{n}{k}$. Therefore $\binom{n}{k}$ is often read as "n choose k". The properties of binomial coefficients have led to extending the meaning of the symbol $\binom{n}{k}$ beyond the basic case where n and k are nonnegative integers with $k \leq n$; such expressions are then still called binomial coefficients.

The notation $\binom{n}{k}$ was introduced by Andreas von Ettingshausen in 1826, although the numbers were already known centuries before that. The earliest known detailed discussion of binomial coefficients is in a tenth-century commentary, by Halayudha, on an ancient Sanskrit text, Pingala's Chanda%úβstra. In about 1150, the Indian mathematician Bhaskaracharya gave an exposition of binomial coefficients in his book Lilavati.

Alternative notations include $C(n, k)$, ${}_nC_k$, ${}^nC_k$, C^k_n, C^n_k, $C_{n,k}$ in all of which the C stands for combinations or *choices*. Many calculators use similar variants of the C notation as it can be represented on a single-line display.

For natural numbers (taken to include 0) n and k, the binomial coefficient $\binom{n}{k}$ can be defined as the coefficient of the monomial X^k in the expansion of $(1 + X)^n$. The same coefficient also occurs (if $k \leq n$) in the binomial formula

$$(x+y)^n = \sum_{k=0}^{n} \binom{n}{k} x^{n-k} y^k$$

(valid for any elements x,y of a commutative ring), which explains the name "binomial coefficient".

Another occurrence of this number is in combinatorics, where it gives the number of ways, disregarding order, that k objects can be chosen from among n objects; more formally, the number of k-element subsets (or k-combinations) of an n-element set. This number can be seen as equal to the one of the first definition, independently of any of the formulas below to compute it: if in each of the n factors of the power $(1 + X)^n$ one temporarily labels the term X with an index i (running from 1 to n), then each subset of k indices gives after expansion a contribution X^k, and the coefficient of that monomial in the result will be the number of such subsets. This shows in particular that $\binom{n}{k}$ is a natural number for any natural numbers n and k. There are many other combinatorial interpretations of binomial coefficients (counting problems for which the answer is given by a binomial coefficient expression) $\binom{n}{k}$, for instance the number of words formed of n bits (digits 0 or 1) whose sum is k is given by , while the number of ways to write $k = a_1 + a_2 + \cdots + a_n$ where every a_i is a nonnegative integer is given by $\binom{n+k-1}{n-1}$. Most of these interpretations are easily seen to be equivalent to counting k-combinations.

Computing the Value of Binomial Coefficients

Several methods exist to compute the value of $\binom{n}{k}$ without actually expanding a binomial power or counting k-combinations.

Recursive Formula

One method uses the recursive, purely additive, formula

$$\binom{n}{k} = \binom{n-1}{k-1} + \binom{n-1}{k} \quad \text{for all integers } n, k : 1 \le k \le n-1,$$

with initial/boundary values

$$\binom{n}{0} = \binom{n}{n} = 1 \quad \text{for all integers } n \ge 0,$$

The formula follows from considering the set {1,2,3,...,n} and counting separately (a) the k-element groupings that include a particular set element, say "i", in every group (since "i" is already chosen to fill one spot in every group, we need only choose $k-1$ from the remaining $n-1$) and (b) all the k-groupings that don't include "i"; this enumerates all the possible k-combinations of n elements. It also follows from tracing the contributions to X^k in $(1+X)^{n-1}(1+X)$. As there is zero X^{n+1} or X^{-1} in $(1+X)^n$, one might extend the definition beyond the above boundaries to include $\binom{n}{k} = 0$ when either $k > n$ or $k < 0$. This recursive formula then allows the construction of Pascal's triangle, surrounded by white spaces where the zeros, or the trivial coefficients, would be.

Multiplicative Formula

A more efficient method to compute individual binomial coefficients is given by the formula

$$\binom{n}{k} = \frac{n^{\underline{k}}}{k!} = \frac{n(n-1)(n-2)\cdots(n-(k-1))}{k(k-1)(k-2)\cdots 1} = \prod_{i=1}^{k} \frac{n-(k-i)}{i} = \prod_{i=1}^{k} \frac{n+1-i}{i},$$

where the numerator of the first fraction $n^{\underline{k}}$ is expressed as a falling factorial power. This formula is easiest to understand for the combinatorial interpretation of binomial coefficients. The numerator gives the number of ways to select a sequence of k distinct objects, retaining the order of selection, from a set of n objects. The denominator counts the number of distinct sequences that define the same k-combination when order is disregarded.

Factorial Formula

Finally, though computationally unsuitable, there is the compact form, often used in proofs and derivations, which makes repeated use of the familiar factorial function:

$$\binom{n}{k} = \frac{n!}{k!(n-k)!} \quad \text{for } 0 \le k \le n,$$

where $n!$ denotes the factorial of n. This formula follows from the multiplicative formula above by multiplying numerator and denominator by $(n - k)!$; as a consequence it involves many factors common to numerator and denominator. It is less practical for explicit computation unless common factors are first cancelled (in particular since factorial values grow very rapidly). The formula does exhibit a symmetry that is less evident from the multiplicative formula (though it is from the definitions)

$$\binom{n}{k} = \binom{n}{n-k} \quad \text{for } 0 \le k \le n, \tag{1}$$

which leads to a more efficient multiplicative computational routine. Using the falling factorial notation,

$$\binom{n}{k} = \begin{cases} n^{\underline{k}} / k!, & \text{if } k \le \frac{n}{2} \\ n^{\underline{n-k}} / (n-k)!, & \text{if } k > \frac{n}{2} \end{cases}$$

Generalization and Connection to the Binomial Series

The multiplicative formula allows the definition of binomial coefficients to be extended by replacing n by an arbitrary number α (negative, real, complex) or even an element of any commutative ring in which all positive integers are invertible:

$$\binom{\alpha}{k} = \frac{\alpha^{\underline{k}}}{k!} = \frac{\alpha(\alpha-1)(\alpha-2)\cdots(\alpha-k+1)}{k(k-1)(k-2)\cdots 1} \quad \text{for } k \in \mathbb{N} \text{ and arbitrary } \alpha.$$

With this definition one has a generalization of the binomial formula (with one of the variables set to 1), which justifies still calling the $\binom{\alpha}{k}$ binomial coefficients:

$$(1+X)^{\alpha} = \sum_{k=0}^{\infty} \binom{\alpha}{k} X^{k}. \tag{2}$$

This formula is valid for all complex numbers α and X with $|X| < 1$. It can also be interpreted as an identity of formal power series in X, where it actually can serve as definition of arbitrary powers of series with constant coefficient equal to 1; the point is that with this definition all identities hold that one expects for exponentiation, notably

$$(1+X)^{\alpha}(1+X)^{\beta} = (1+X)^{\alpha+\beta} \quad \text{and} \quad ((1+X)^{\alpha})^{\beta} = (1+X)^{\alpha\beta}.$$

If α is a nonnegative integer n, then all terms with $k > n$ are zero, and the infinite series becomes a finite sum, thereby recovering the binomial formula. However for other values of α, including negative integers and rational numbers, the series is really infinite.

Pascal's Triangle

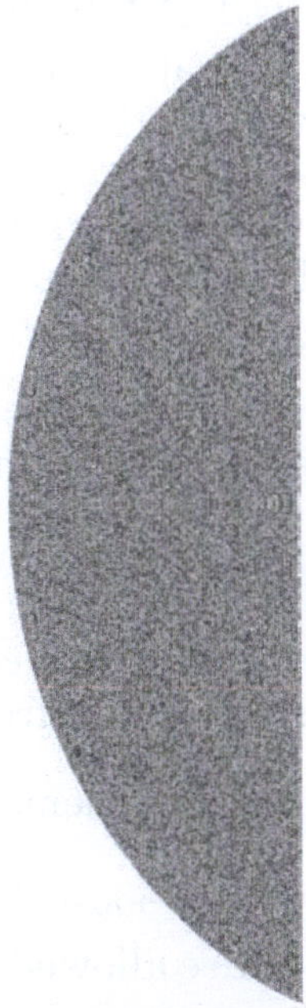

Figure: *1000th row of Pascal's triangle, arranged vertically, with grey-scale representations of decimal digits of the coefficients, right-aligned. The left boundary of the image corresponds roughly to the graph of the logarithm of the binomial coefficients, and illustrates that they form a log-concave sequence.*

Pascal's rule is the important recurrence relation:

$$\binom{n}{k}+\binom{n}{k+1}=\binom{n+1}{k+1}, \tag{3}$$

which can be used to prove by mathematical induction that $\binom{n}{k}$ is a natural number for all n and k, (equivalent to the statement that k! divides the product of k consecutive integers), a fact that is not immediately obvious from formula (1).

Pascal's rule also gives rise to Pascal's triangle:

0:									1								
1:								1		1							
2:							1		2		1						
3:						1		3		3		1					
4:					1		4		6		4		1				
5:				1		5		10		10		5		1			
6:			1		6		15		20		15		6		1		
7:		1		7		21		35		35		21		7		1	
8:	1		8		28		56		70		56		28		8		1

Row number n contains the numbers $\binom{n}{k}$ for $k = 0,\ldots,n$. It is constructed by starting with ones at the outside and then always adding two adjacent numbers and writing the sum directly underneath. This method allows the quick calculation of binomial coefficients without the need for fractions or multiplications. For instance, by looking at row number 5 of the triangle, one can quickly read off that

$$(x + y)^5 = 1\,x^5 + 5\,x^4y + 10\,x^3y^2 + 10\,x^2y^3 + 5\,x\,y^4 + 1\,y^5.$$

The differences between elements on other diagonals are the elements in the previous diagonal, as a consequence of the recurrence relation (3) above.

Combinatorics and Statistics

Binomial coefficients are of importance in combinatorics, because they provide ready formulas for certain frequent counting problems:

- There are $\binom{n}{k}$ ways to choose k elements from a set of n elements.
- There are $\binom{n+k-1}{k}$ ways to choose k elements from a set of n elements if repetitions are allowed.
- There are $\binom{n+k}{k}$ strings containing k ones and n zeros.
- There are $\binom{n+1}{k}$ strings consisting of k ones and n zeros such that no two ones are adjacent.

- The Catalan numbers are $\frac{1}{n+1}\binom{2n}{n}$.
- The binomial distribution in statistics is $\binom{n}{k}p^k(1-p)^{n-k}$.
- The formula for a Bézier curve.

Binomial Coefficients as Polynomials

For any nonnegative integer *k*, the expression $\binom{t}{k}$ can be simplified and defined as a polynomial divided by *k*!:

$$\binom{t}{k}=\frac{(t)_k}{k!}=\frac{(t)_k}{(k)_k}=\frac{t(t-1)(t-2)\cdots(t-k+1)}{k(k-1)(k-2)\cdots 2\cdot 1};$$

This presents a polynomial in *t* with rational coefficients.

As such, it can be evaluated at any real or complex number *t* to define binomial coefficients with such first arguments. These "generalized binomial coefficients" appear in Newton's generalized binomial theorem.

For each *k*, the polynomial $\binom{t}{k}$ can be characterized as the unique degree *k* polynomial *p*(*t*) satisfying $p(0) = p(1) = \ldots = p(k-1) = 0$ and $p(k) = 1$.

Its coefficients are expressible in terms of Stirling numbers of the first kind:

$$\binom{t}{k}=\sum_{i=0}^{k}\frac{s_{k,i}}{k!}t^i.$$

The derivative of $\binom{t}{k}$ can be calculated by logarithmic differentiation:

$$\frac{\mathrm{d}}{\mathrm{d}t}\binom{t}{k}=\binom{t}{k}\sum_{i=0}^{k-1}\frac{1}{t-i}.$$

Binomial Coefficients as a Basis for the Space of Polynomials

Over any field of characteristic 0 (that is, any field that contains the rational numbers), each polynomial *p*(*t*) of degree at most *d* is uniquely expressible as a linear combination $\sum_{k=0}^{d}a_k\binom{t}{k}$ of binomial coefficients. The coefficient a_k is the kth difference of the sequence *p*(0), *p*(1), ..., *p*(*k*). Explicitly,

$$a_k = \sum_{i=0}^{k} (-1)^{k-i} \binom{k}{i} p(i). \tag{4}$$

Integer-Valued Polynomials

Each polynomial $\binom{t}{k}$ is integer-valued: it takes integer values at integer inputs. (One way to prove this is by induction on k, using Pascal's identity.) Therefore any integer linear combination of binomial coefficient polynomials is integer-valued too.

Conversely, (4) shows that any integer-valued polynomial is an integer linear combination of these binomial coefficient polynomials. More generally, for any subring R of a characteristic 0 field K, a polynomial in $K[t]$ takes values in R at all integers if and only if it is an R-linear combination of binomial coefficient polynomials.

Example

The integer-valued polynomial $3t(3t + 1)/2$ can be rewritten as

$$9\binom{t}{2} + 6\binom{t}{1} + 0\binom{t}{0}.$$

Identities Involving Binomial Coefficients

The factorial formula facilitates relating nearby binomial coefficients. For instance, if k is a positive integer and n is arbitrary, then

$$\binom{n}{k} = \frac{n}{k}\binom{n-1}{k-1} \tag{5}$$

and, with a little more work,

$$\binom{n-1}{k} - \binom{n-1}{k-1} = \frac{n-2k}{n}\binom{n}{k}.$$

Moreover, the following may be useful:

$$\binom{n}{h}\binom{n-h}{k} = \binom{n}{k}\binom{n-k}{h}.$$

For constant n, we have the following recurrence:

$$\binom{n}{k} = \frac{n+1-k}{k}\binom{n}{k-1}.$$

Series Involving Binomial Coefficients

The formula $\sum_{k=0}^{n}\binom{n}{k} = 2^n$ is obtained from (2) using $x = 1$. This is equivalent to saying that the elements in one row of Pascal's triangle always add up to two raised to an integer power. A combinatorial interpretation of this fact involving double counting is given by counting subsets of size 0, size 1, size 2, and so on up to size n of a set S of n elements. Since we count the number of subsets of size i for $0 \le i \le n$, this sum must be equal to the number of subsets of S, which is known to be 2^n. That is, Equation 5 is a statement that the power set for a finite set with n elements has size 2^n. More explicitly, consider a bit string with n digits. This bit string can be used to represent 2^n numbers. Now consider all of the bit strings with no ones in them. There is just one, or rather n choose 0. Next consider the number of bit strings with just a single one in them. There are n, or rather n choose 1. Continuing this way we can see that the equation above holds.

The formulas

$$\sum_{k=0}^{n} k\binom{n}{k} = n2^{n-1} \tag{6}$$

and $\sum_{k=0}^{n} k^2\binom{n}{k} = (n+n^2)2^{n-2}$ follow from (2) after differentiating with respect to x (twice in the latter) and then substituting $x = 1$.

The Chu–Vandermonde identity, which holds for any complex-values m and n and any non-negative integer k, is

$$\sum_{j=0}^{k}\binom{m}{j}\binom{n-m}{k-j} = \binom{n}{k} \tag{7}$$

and can be found by examination of the coefficient of x^k in the expansion of $(1 + x)^m (1 + x)^{n-m} = (1 + x)^n$ using equation (2). When $m = 1$, equation (7) reduces to equation (3).

A similar looking formula, which applies for any integers j, k, and n satisfying $0 \le j \le k \le n$, is

$$\sum_{m=0}^{n}\binom{m}{j}\binom{n-m}{k-j} = \binom{n+1}{k+1}, \tag{8}$$

and can be found by examination of the coefficient of x^{n+1} in the expansion of $x\left(\frac{x^j}{(1-x)^{j+1}}\right)\left(\frac{x^{k-j}}{(1-x)^{k-j+1}}\right) = \frac{x^{k+1}}{(1-x)^{k+2}}$ using $\frac{x^l}{(1-x)^{l+1}} = \sum_{p=0}^{\infty}\binom{p}{l}x^p$. When $j = k$, equation (8) gives

$$\sum_{m=0}^{n}\binom{m}{k}=\binom{n+1}{k+1}.$$

From expansion (7) using $n = 2m$, $k = m$, and (1), one finds

$$\sum_{j=0}^{m}\binom{m}{j}^2=\binom{2m}{m}. \tag{9}$$

Let $F(n)$ denote the n-th Fibonacci number. We obtain a formula about the diagonals of Pascal's triangle $\sum_{k=0}^{\left[\frac{n}{2}\right]}\binom{n-k}{k}=F(n+1)$.

This can be proved by induction using (3) or by Zeckendorf's representation (Just note that the lhs gives the number of subsets of $\{F(2),\ldots,F(n)\}$ without consecutive members, which also form all the numbers below $F(n + 1)$). A combinatorial proof is given below.

Also using (3) and induction, one can show that $\sum_{j=k}^{n}\binom{j}{k}=\binom{n+1}{k+1}$.

Although there is no closed formula for

$$\sum_{j=0}^{k}\binom{n}{j}$$

(unless one resorts to Hypergeometric functions), one can again use (3) and induction, to show that for $k = 0, \ldots, n - 1$ $\sum_{j=0}^{k}(-1)^j\binom{n}{j}=(-1)^k\binom{n-1}{k}$

as well as $\sum_{j=0}^{n}(-1)^j\binom{n}{j}=0$

[except in the trivial case where $n = 0$, where the result is 1 instead] which is itself a special case of the result from the theory of finite differences that for any polynomial $P(x)$ of degree less than n, $\sum_{j=0}^{n}(-1)^j\binom{n}{j}P(j)=0$. Differentiating (2) k times and setting $x = -1$ yields this for $P(x)=x(x-1)\cdots(x-k+1)$, when $0 \le k < n$, and the general case follows by taking linear combinations of these.

When $P(x)$ is of degree less than or equal to n,

$$\sum_{j=0}^{n}(-1)^j\binom{n}{j}P(n-j)=n!a_n \tag{10}$$

where a_n is the coefficient of degree n in $P(x)$.

More generally for (10),

$\sum_{j=0}^{n}(-1)^j\binom{n}{j}P(m+(n-j)d)=d^n n! a_n$ where m and d are complex numbers. This follows immediately applying (10) to the polynomial $Q(x):=P(m+dx)$ instead of $P(x)$, and observing that $Q(x)$ has still degree less than or equal to n, and that its coefficient of degree n is $d^n a_n$.

The infinite series $\sum_{j=m}^{\infty}\frac{1}{\binom{j}{k}}=\frac{k}{(k-1)\binom{m-1}{k-1}}$ is convergent for k e– 2. This formula is used in the analysis of the German tank problem. It is equivalent to the formula for the finite sum

$$\sum_{j=m}^{M-1}\frac{1}{k\binom{j}{k}}=\frac{1}{(k-1)\binom{m-1}{k-1}}-\frac{1}{(k-1)\binom{M-1}{k-1}}$$

which is proved for $M>m$ by induction on M.

Using (9) one can derive

$$\sum_{i=0}^{n} i\binom{n}{i}^2=\frac{n}{2}\binom{2n}{n} \text{ and } \sum_{i=0}^{n} i^2\binom{n}{i}^2=n^2\binom{2n-2}{n-1}.$$

Series multisection gives the following identity for the sum of binomial coefficients taken with a step s and offset t $(0\leqslant t<s)$ as a closed-form sum of s terms:

$$\binom{n}{t}+\binom{n}{t+s}+\binom{n}{t+2s}+\ldots=\frac{1}{s}\sum_{j=0}^{s-1}\left(2\cos\frac{\pi j}{s}\right)^n\cos\frac{\pi(n-2t)j}{s}.$$

Identities with Combinatorial Proofs

Many identities involving binomial coefficients can be proved by combinatorial means. For example, the following identity for nonnegative integers $n \geq q$ (which reduces to (6) when $q = 1$):

$$\sum_{k=q}^{n}\binom{n}{k}\binom{k}{q}=2^{n-q}\binom{n}{q}$$

can be given a double counting proof as follows. The left side counts the number of ways of selecting a subset of $[n] = \{1, 2, ..., n\}$ with at least q elements, and marking q elements among those selected. The right side counts the same parameter, because there are $\binom{n}{q}$ ways of choosing a set of q marks and they occur in all subsets that additionally contain some subset of the remaining elements, of which there are 2^{n-q}.

In the Pascal's rule

$$\binom{n}{k}=\binom{n-1}{k-1}+\binom{n-1}{k}$$

both sides count the number of k-element subsets of $[n]$ with the right hand side first grouping them into those that contain element n and those that do not.

The identity (9) also has a combinatorial proof. The identity reads

$$\sum_{k=0}^{n}\binom{n}{k}^2=\binom{2n}{n}.$$

Suppose you have $2n$ empty squares arranged in a row and you want to mark (select) n of them. There are $\binom{2n}{n}$ ways to do this. On the other hand, you may select your n squares by selecting k squares from among the first n and $n-k$ squares from the remaining n squares; any k from 0 to n will work. This gives

$$\sum_{k=0}^{n}\binom{n}{k}\binom{n}{n-k}=\binom{2n}{n}.$$

Now apply (5) to get the result.

The identity (9),

$$\sum_{k=0}^{\left[\frac{n}{2}\right]}\binom{n-k}{k}=F(n+1)$$

has the following combinatorial proof. The number $\binom{n-k}{k}$ denotes the number of paths in a two-dimensional lattice from $(0,0)$ to $(k,n-k)$ using steps $(0,1)$ and $(1,1)$. This is easy to see: there are $(n-k)$ steps in total and one may choose the $k(0,1)$ steps. Now, replace each $(1,1)$ step by a $(0,2)$ step; note that there are exactly k. Then one arrives at point $(0,n)$ using steps $(0,1)$ and $(0,2)$. Doing this for all k between 0 and $\left[\frac{n}{2}\right]$ gives all paths from $(0,0)$ to $(0,n)$ using steps $(0,1)$ and $(0,2)$. Clearly, there are exactly $F(n+1)$ such paths.

Sum of Coefficients Row

The number of k-combinations for all k, $\sum_{0\le k\le n}\binom{n}{k}=2^n$, is the sum of the nth row (counting from 0) of the binomial coefficients.

These combinations are enumerated by the 1 digits of the set of base 2 numbers counting from 0 to 2^n-1, where each digit position is an item from the set of n.

Dixon's Identity

Dixon's identity is

$$\sum_{k=-a}^{a}(-1)^k\binom{2a}{k+a}^3=\frac{(3a)!}{(a!)^3}$$

or, more generally,

$$\sum_{k=-a}^{a}(-1)^k\binom{a+b}{a+k}\binom{b+c}{b+k}\binom{c+a}{c+k}=\frac{(a+b+c)!}{a!b!c!},$$

where a, b, and c are non-negative integers.

Continuous Identities

Certain trigonometric integrals have values expressible in terms of binomial coefficients:

For $m,n\in\mathbb{N}$

$$\int_{-\pi}^{\pi}\cos((2m-n)x)\cos^n x\,dx=\frac{\pi}{2^{n-1}}\binom{n}{m}$$

$$\int_{-\pi}^{\pi}\sin((2m-n)x)\sin^n x\,dx=\begin{cases}(-1)^{m+(n+1)/2}\dfrac{\pi}{2^{n-1}}\dbinom{n}{m} & n \text{ odd}\\ 0 & \text{otherwise}\end{cases}$$

$$\int_{-\pi}^{\pi}\cos((2m-n)x)\sin^n x\,dx=\begin{cases}(-1)^{m+(n+1)/2}\dfrac{\pi}{2^{n-1}}\dbinom{n}{m} & n \text{ even}\\ 0 & \text{otherwise}\end{cases}$$

These can be proved by using Euler's formula to convert trigonometric functions to complex exponentials, expanding using the binomial theorem, and integrating term by term.

Generating Functions

Ordinary Generating Functions: For a fixed n, the ordinary generating function of the sequence $\binom{n}{0}, \binom{n}{1}, \binom{n}{2}, \cdots$ is:

$$\sum_k \binom{n}{k} x^k = (1+x)^n.$$

For a fixed k, the ordinary generating function of the sequence $\binom{0}{k}, \binom{1}{k}, \binom{2}{k}, \cdots$ is:

$$\sum_{n=k}^{\infty} \binom{n}{k} y^n = \frac{y^k}{(1-y)^{k+1}}.$$

The bivariate generating function of the binomial coefficients is:

$$\sum_{n,k} \binom{n}{k} x^k y^n = \frac{1}{1-y-xy}.$$

Another bivariate generating function of the binomial coefficients, which is symmetric, is:

$$\sum_{n,k} \binom{n+k}{k} x^k y^n = \frac{1}{1-x-y}.$$

Exponential Generating Function

The exponential bivariate generating function of the binomial coefficients is:

$$\sum_{n,k} \frac{1}{(n+k)!} \binom{n+k}{k} x^k y^n = e^{x+y}.$$

Divisibility Properties

In 1852, Kummer proved that if m and n are nonnegative integers and p is a prime number, then the largest power of p dividing $\binom{m+n}{m}$ equals p^c, where c is the number of carries when m and n are added in base p. Equivalently, the exponent of a prime p in $\binom{n}{k}$ equals the number of nonnegative integers j such that the fractional part of k/p^j is greater than the fractional part of n/p^j. It can be deduced from this that $\binom{n}{k}$ is divisible by $n/\gcd(n,k)$. In particular therefore it follows that

p divides $\binom{p^r}{s}$ for all positive integers r and s such that $s < p^r$. However this is not true of higher powers of p: for example 9 does not divide $\binom{9}{6}$.

A somewhat surprising result by David Singmaster (1974) is that any integer divides almost all binomial coefficients. More precisely, fix an integer d and let $f(N)$ denote the number of binomial coefficients $\binom{n}{k}$ with $n < N$ such that d divides $\binom{n}{k}$. Then

$$\lim_{N\to\infty} \frac{f(N)}{N(N+1)/2} = 1.$$

Since the number of binomial coefficients $\binom{n}{k}$ with $n < N$ is $N(N + 1) / 2$, this implies that the density of binomial coefficients divisible by d goes to 1.

Another fact: An integer n e– 2 is prime if and only if all the intermediate binomial coefficients

$$\binom{n}{1}, \binom{n}{2}, \ldots, \binom{n}{n-1}$$

are divisible by n.

Proof: When p is prime, p divides

for all $0 < k < p$

because it is a natural number and the numerator has a prime factor p but the denominator does not have a prime factor p.

When n is composite, let p be the smallest prime factor of n and let k = n/p. Then $0 < p < n$ and

$$\binom{p}{k} = \frac{p\cdot(p-1)\cdots(p-k+1)}{k\cdot(k-1)\cdots 1}$$

$$\binom{n}{p} = \frac{n(n-1)(n-2)\cdots(n-p+1)}{p!} = \frac{k(n-1)(n-2)\cdots(n-p+1)}{(p-1)!} \not\equiv 0 \pmod{n}$$

otherwise the numerator $k(n - 1)(n - 2)\times\ldots\times(n - p + 1)$ has to be divisible by $n = k\times p$, this can only be the case when $(n - 1)(n - 2)\times\ldots\times(n - p + 1)$ is divisible by p. But n is divisible by p, so p does not divide $n - 1, n - 2, \ldots, n - p + 1$ and because p is prime, we know that p does not divide $(n - 1)(n - 2)\times\ldots\times(n - p + 1)$ and so the numerator cannot be divisible by n.

Bounds and Asymptotic Formulas

The following bounds for $\binom{n}{k}$ hold:

$$\left(\frac{n}{k}\right)^k \le \binom{n}{k} \le \frac{n^k}{k!} \le \left(\frac{n\cdot e}{k}\right)^k \text{ for } 1 \le k \le n.$$

Stirling's approximation yields the bounds:

$\sqrt{n}\binom{2n}{n} \ge 2^{2n-1}$ and, in general, $\sqrt{n}\binom{mn}{n} \ge \dfrac{m^{m(n-1)+1}}{(m-1)^{(m-1)(n-1)}}$ for $m \ge 2$ and $n \ge 1$,

and the approximation

$$\binom{2n}{n} \sim \frac{4^n}{\sqrt{\pi n}} \text{ as } n \to \infty.$$

For both n and k much larger than 1, Stirling's approximation also yields the following asymptotic approximation:

$$\log\binom{n}{k} \sim nH\left(\frac{k}{n}\right)$$

Where $H(\epsilon) = -\epsilon\log(\epsilon) - (1-\epsilon)\log(1-\epsilon)$ is the binary entropy of ϵ.

When is large and is much smaller than , one can also write

$$\binom{n}{k} = \frac{n(n-1)\ldots(n-k+1)}{k!} \approx \frac{(n-k/2)^k}{k^k e^{-k}\sqrt{2\pi k}} = \frac{(n/k-0.5)^k e^k}{\sqrt{2\pi k}}$$

and therefore

$$\log\binom{n}{k} \approx k\ln(n/k-0.5) + k - 0.5\ln(2\pi k)$$

If more precision is desired, one can $\ln(n(n-1)\ldots(n-k+1))$ approximate with an integral, obtaining

$$\log\binom{n}{k} \approx (n+0.5)\ln\frac{n+0.5}{n-k+0.5} + k\ln\frac{n-k+0.5}{k} - 0.5\ln(2\pi k)$$

For $n = 20$ and $k = 10$, $\log\binom{n}{k} \approx 12.127$, and these approximations yield 12.312 and 12.133 respectively.

The infinite product formula (cf. Gamma function, alternative definition)

$$(-1)^k\binom{z}{k}=\binom{-z+k-1}{k}=\frac{1}{\Gamma(-z)}\frac{1}{(k+1)^{z+1}}\prod_{j=k+1}\frac{(1+\frac{1}{j})^{-z-1}}{1-\frac{z+1}{j}}$$

yields the asymptotic formulas

$$\binom{z}{k}\approx\frac{(-1)^k}{\Gamma(-z)k^{z+1}} \quad \text{and} \quad \binom{z+k}{k}=\frac{k^z}{\Gamma(z+1)}\left(1+\frac{z(z+1)}{2k}+\mathcal{O}\left(k^{-2}\right)\right)$$

as $k\to\infty$.

This asymptotic behaviour is contained in the approximation

$$\binom{z+k}{k}\approx\frac{e^{z(H_k-\gamma)}}{\Gamma(z+1)}$$

as well. (Here H_k is the k-th harmonic number and γ is the Euler–Mascheroni constant.)

The sum of binomial coefficients can be bounded by a term exponential in n and the binary entropy of the largest n/k that occurs. More precisely, for $n\geq 1$ and $0<\epsilon<\frac{1}{2}$, it holds

$$\sum_{k=0}^{[\epsilon n]}\binom{n}{k}\leq 2^{H(\epsilon)\cdot n},$$

where $H(\epsilon)=-\epsilon\cdot\log\epsilon-(1-\epsilon)\cdot\log(1-\epsilon)$ is the binary entropy of ϵ.

A simple and rough upper bound for the sum of binomial coefficients is given by the formula below (not difficult to prove)

$$\sum_{i=0}^{k}\binom{n}{i}\leq 1+n^k$$

Generalization to Multinomials

Binomial coefficients can be generalized to multinomial coefficients defined to be the number:

$$\binom{n}{k_1,k_2,\ldots,k_r}=\frac{n!}{k_1!k_2!\cdots k_r!}$$

where

$$\sum_{i=1}^{r}k_i=n.$$

While the binomial coefficients represent the coefficients of $(x+y)^n$, the multinomial coefficients represent the coefficients of the polynomial

$$(x_1 + x_2 + \cdots + x_r)^n.$$

The case $r = 2$ gives binomial coefficients:

$$\binom{n}{k_1, k_2} = \binom{n}{k_1, n-k_1} = \binom{n}{k_1} = \binom{n}{k_2}.$$

The combinatorial interpretation of multinomial coefficients is distribution of n distinguishable elements over r (distinguishable) containers, each containing exactly k_i elements, where i is the index of the container.

Multinomial coefficients have many properties similar to these of binomial coefficients, for example the recurrence relation:

$$\binom{n}{k_1, k_2, \ldots, k_r} = \binom{n-1}{k_1 - 1, k_2, \ldots, k_r} + \binom{n-1}{k_1, k_2 - 1, \ldots, k_r} + \ldots + \binom{n-1}{k_1, k_2, \ldots, k_r - 1}$$

and symmetry:

$$\binom{n}{k_1, k_2, \ldots, k_r} = \binom{n}{k_{\sigma_1}, k_{\sigma_2}, \ldots, k_{\sigma_r}}$$

where (σ_i) is a permutation of $(1,2,\ldots,r)$.

Taylor Series

Using Stirling numbers of the first kind the series expansion around any arbitrarily chosen point z_0 is

$$\binom{z}{k} = \frac{1}{k!}\sum_{i=0}^{k} z^i s_{k,i} = \sum_{i=0}^{k}(z - z_0)^i \sum_{j=i}^{k}\binom{z_0}{j-i}\frac{s_{k+i-j,i}}{(k+i-j)!}$$

$$= \sum_{i=0}^{k}(z - z_0)^i \sum_{j=i}^{k} z_0^{j-i}\binom{j}{i}\frac{s_{k,j}}{k!}.$$

Binomial Coefficient with n = ½

The definition of the binomial coefficients can be extended to the case where n is real and k is integer.

In particular, the following identity holds for any non-negative integer :

$$\binom{1/2}{k} = \binom{2k}{k}\frac{(-1)^{k+1}}{2^{2k}(2k-1)}.$$

This shows up when expanding $\sqrt{1+x}$ into a power series using the Newton binomial series :

$$\sqrt{1+x} = \sum_{k\geqslant 0}\binom{1/2}{k}x^k.$$

Identity for the Product of Binomial Coefficients

One can express the product of binomial coefficients as a linear combination of binomial coefficients:

$$\binom{z}{m}\binom{z}{n} = \sum_{k=0}^{m}\binom{m+n-k}{k,m-k,n-k}\binom{z}{m+n-k}$$

where the connection coefficients are multinomial coefficients. In terms of labelled combinatorial objects, the connection coefficients represent the number of ways to assign m+n-k labels to a pair of labelled combinatorial objects—of weight m and n respectively—that have had their first k labels identified, or glued together to get a new labelled combinatorial object of weight m+n-k. (That is, to separate the labels into three portions to apply to the glued part, the unglued part of the first object, and the unglued part of the second object.) In this regard, binomial coefficients are to exponential generating series what falling factorials are to ordinary generating series.

Partial Fraction Decomposition

The partial fraction decomposition of the reciprocal is given by

$$\frac{1}{\binom{z}{n}} = \sum_{i=0}^{n-1}(-1)^{n-1-i}\binom{n}{i}\frac{n-i}{z-i}, \quad \text{and} \quad \frac{1}{\binom{z+n}{n}} = \sum_{i=1}^{n}(-1)^{i-1}\binom{n}{i}\frac{i}{z+i}.$$

Newton's Binomial Series

Newton's binomial series, named after Sir Isaac Newton, is a generalization of the binomial theorem to infinite series:

$$(1+z)^{\alpha} = \sum_{n=0}^{\infty}\binom{\alpha}{n}z^n = 1+\binom{\alpha}{1}z+\binom{\alpha}{2}z^2+\cdots.$$

The identity can be obtained by showing that both sides satisfy the differential equation $(1 + z) f'(z) = \alpha f(z)$. The radius of convergence of this series is 1. An alternative expression is

$$\frac{1}{(1-z)^{\alpha+1}} = \sum_{n=0}^{\infty}\binom{n+\alpha}{n}z^n$$

where the identity

$$\binom{n}{k} = (-1)^k \binom{k-n-1}{k}$$

is applied.

Multiset (Rising) Binomial Coefficient

Binomial coefficients count subsets of prescribed size from a given set. A related combinatorial problem is to count multisets of prescribed size with elements drawn from a given set, that is, to count the number of ways to select a certain number of elements from a given set with the possibility of selecting the same element repeatedly. The resulting numbers are called multiset coefficients; the number of ways to "multichoose" (i.e., choose with replacement) k items from an n element set is denoted $\left(\!\binom{n}{k}\!\right)$.

To avoid ambiguity and confusion with n's main denotation in this article,

let $f = n = r + (k - 1)$ and $r = f - (k - 1)$.

Multiset coefficients may be expressed in terms of binomial coefficients by the rule

$$\binom{f}{k} = \left(\!\binom{r}{k}\!\right) = \binom{r+k-1}{k}.$$

One possible alternative characterization of this identity is as follows: We may define the falling factorial as

$$(f)_k = f^{\underline{k}} = (f-k+1)\cdots(f-3)\cdot(f-2)\cdot(f-1)\cdot f,$$

and the corresponding rising factorial as

$$r^{(k)} = r^{\overline{k}} = r\cdot(r+1)\cdot(r+2)\cdot(r+3)\cdots(r+k-1);$$

so, for example,

$$17\cdot18\cdot19\cdot20\cdot21 = (21)_5 = 21^{\underline{5}} = 17^{\overline{5}} = 17^{(5)}.$$

Then the binomial coefficients may be written as

$$\binom{f}{k} = \frac{(f)_k}{k!} = \frac{(f-k+1)\cdots(f-2)\cdot(f-1)\cdot f}{1\cdot2\cdot3\cdot4\cdot5\cdots k},$$

while the corresponding multiset coefficient is defined by replacing the falling with the rising factorial:

$$\left(\!\!\binom{r}{k}\!\!\right)=\frac{r^{(k)}}{k!}=\frac{r\cdot(r+1)\cdot(r+2)\cdots(r+k-1)}{1\cdot2\cdot3\cdot4\cdot5\cdots k}.$$

Generalization to Negative Integers

For any n,

$$\binom{-n}{k}=\frac{-n\cdot-(n+1)\ldots-(n+k-2)\cdot-(n+k-1)}{k!}$$

$$=(-1)^k\,\frac{n\cdot(n+1)\cdot(n+2)\cdots(n+k-1)}{k!}$$

$$=(-1)^k\binom{n+k-1}{k}$$

$$=(-1)^k\left(\!\!\binom{n}{k}\!\!\right).$$

In particular, binomial coefficients evaluated at negative integers are given by signed multiset coefficients. In the special case $n=-1$, this reduces to $(-1)^k=\binom{-1}{k}=\left(\!\!\binom{-k}{k}\!\!\right)$.

For example, if n = -4 and k = 7, then r = 4 and f = 10:

$$\binom{-4}{7}=\frac{-10\cdot-9\cdot-8\cdot-7\cdot-6\cdot-5\cdot-4}{1\cdot2\cdot3\cdot4\cdot5\cdot6\cdot7}$$

$$=(-1)^7\,\frac{4\cdot5\cdot6\cdot7\cdot8\cdot9\cdot10}{1\cdot2\cdot3\cdot4\cdot5\cdot6\cdot7}$$

$$=\left(\!\!\binom{-7}{7}\!\!\right)\left(\!\!\binom{4}{7}\!\!\right)=\binom{-1}{7}\binom{10}{7}.$$

Two Real or Complex Valued Arguments

The binomial coefficient is generalized to two real or complex valued arguments using the gamma function or beta function via

$$\binom{x}{y}=\frac{\Gamma(x+1)}{\Gamma(y+1)\Gamma(x-y+1)}=\frac{1}{(x+1)\mathrm{B}(x-y+1,y+1)}.$$

This definition inherits these following additional properties from Γ:

$$\binom{x}{y} = \frac{\sin(y\pi)}{\sin(x\pi)}\binom{-y-1}{-x-1} = \frac{\sin((x-y)\pi)}{\sin(x\pi)}\binom{y-x-1}{y};$$

moreover,

$$\binom{x}{y} \cdot \binom{y}{x} = \frac{\sin((x-y)\pi)}{(x-y)\pi}.$$

The resulting function has been little-studied, apparently first being graphed in (Fowler 1996). Notably, many binomial identities fail: $\binom{n}{m} = \binom{n}{n-m}$ but $\binom{-n}{m} \neq \binom{-n}{-n-m}$ for n positive (so $-n$ negative). The behaviour is quite complex, and markedly different in various octants (that is, with respect to the x and y axes and the line $y = x$), with the behaviour for negative x having singularities at negative integer values and a checkerboard of positive and negative regions:

- in the octant $0 \le y \le x$ it is a smoothly interpolated form of the usual binomial, with a ridge ("Pascal's ridge").
- in the octant $0 \le x \le y$ and in the quadrant $x \ge 0, y \le 0$ the function is close to zero.
- in the quadrant $x \le 0, y \ge 0$ the function is alternatingly very large positive and negative on the parallelograms with vertices $(-n, m+1), (-n, m), (-n-1, m-1), (-n-1, m)$
- in the octant $0 > x > y$ the behaviour is again alternatively very large positive and negative, but on a square grid.
- in the octant $-1 > y > x+1$ it is close to zero, except for near the singularities.

Generalization to q-Series

The binomial coefficient has a q-analog generalization known as the Gaussian binomial coefficient.

Generalization to Infinite Cardinals

The definition of the binomial coefficient can be generalized to infinite cardinals by defining:

$$\binom{\alpha}{\beta} = |\{B \subseteq A : |B| = \beta\}|$$

where A is some set with cardinality α. One can show that the generalized binomial coefficient is well-defined, in the sense that no matter what set we choose to represent the cardinal number α, $\binom{\alpha}{\beta}$ will remain the same. For finite cardinals, this definition coincides with the standard definition of the binomial coefficient.

Assuming the Axiom of Choice, one can show that $\binom{\alpha}{\alpha} = 2^{\alpha}$ for any infinite cardinal α.

Binomial Coefficient in Programming Languages

The notation $\binom{n}{k}$ is convenient in handwriting but inconvenient for typewriters and computer terminals. Many programming languages do not offer a standard subroutine for computing the binomial coefficient, but for example both the APL programming language and the (related) J programming language use the exclamation mark: k ! n.

Naive implementations of the factorial formula, such as the following snippet in Python:

```
from math import factorial
def binomialCoefficient(n, k):
    return factorial(n) // (factorial(k) * factorial(n - k))
```

are very slow and are useless for calculating factorials of very high numbers (in languages such as C or Java they suffer from overflow errors because of this reason). A direct implementation of the multiplicative formula works well:

```
def binomialCoefficient(n, k):
    if k < 0 or k > n:
        return 0
    if k == 0 or k == n:
        return 1
    k = min(k, n - k) # take advantage of symmetry
    c = 1
    for i in range(k):
        c = c * (n - i) / (i + 1)
    return c
```

(In Python, range(k) produces a list from 0 to k-1.)

Pascal's rule provides a recursive definition which can also be implemented in Python, although it is less efficient:

```
def binomialCoefficient(n, k):
    if k < 0 or k > n:
        return 0
    if k > n - k: # take advantage of symmetry
        k = n - k
    if k == 0 or n <= 1:
     return 1
        return binomialCoefficient(n-1, k) + binomialCoefficient(n-1, k-1)
```

The example mentioned above can be also written in functional style. The following Scheme example uses the recursive definition

$$\binom{n}{k+1} = \frac{n-k}{k+1}\binom{n}{k}$$

Rational arithmetic can be easily avoided using integer division

$$\binom{n}{k+1} = \left[(n-k)\binom{n}{k}\right] \div (k+1)$$

The following implementation uses all these ideas

```
(define (binomial n k)
;; Helper function to compute C(n,k) via forward recursion
  (define (binomial-iter n k i prev)
    (if (>= i k)
      prev
      (binomial-iter n k (+ i 1) (/ (* (- n i) prev) (+ i 1)))))
;; Use symmetry property C(n,k)=C(n, n-k)
  (if (< k (-  n k))
    (binomial-iter n k 0 1)
    (binomial-iter n (- n k) 0 1)))
```

Another way to compute the binomial coefficient when using large numbers is to recognise that

$$\binom{n}{k} = \frac{n!}{k!(n-k)!} = \frac{\Gamma(n+1)}{\Gamma(k+1)\Gamma(n-k+1)} = \exp(\ln\Gamma(n+1) - \ln\Gamma(k+1) - \ln\Gamma(n-k+1)),$$

where $\ln\Gamma(n)$ denotes the natural logarithm of the gamma function at n. It is a special function that is easily computed and is standard in some programming languages such as using *log_gamma* in Maxima, *LogGamma* in Mathematica, *gammaln* in MATLAB, or *lgamma* in R. Roundoff error may cause the returned value to not be an integer.

Chapter 4

Sequence and Series

In mathematics, a sequence is a list of objects (or events) which have been ordered in a sequential fashion; such that each member either comes before, or after, every other member. More formally, a sequence is a function with a domain equal to the set of positive integers.

A series is a sum of a sequence of terms. That is, a series is a list of numbers with addition operations between them.

A sequence is an ordered list. Like a set, it contains members (also called *elements*, or *terms*). The number of ordered elements (possibly infinite) is called the *length* of the sequence. Unlike a set, order matters, and exactly the same elements can appear multiple times at different positions in the sequence. Most precisely, a sequence can be defined as a function whose domain is a countable totally ordered set, such as the natural numbers.

For example, (M, A, R, Y) is a sequence of letters with the letter 'M' first and 'Y' last. This sequence differs from (A, R, M, Y). Also, the sequence (1, 1, 2, 3, 5, 8), which contains the number 1 at two different positions, is a valid sequence. Sequences can be finite, as in this example, or infinite, such as the sequence of all even positive integers (2, 4, 6,...). Finite sequences are sometimes known as *strings* or *words* and infinite sequences as *streams*. The empty sequence ( ) is included in most notions of sequence, but may be excluded depending on the context.

Examples and Notation

A sequence can be thought of as a list of elements with a particular order. Sequences are useful in a number of mathematical disciplines for studying functions, spaces, and other mathematical structures using the convergence properties of sequences. In particular, sequences are the basis for series, which are important in differential equations and analysis. Sequences are also of interest in their own right and can be studied as patterns or puzzles, such as in the study of prime numbers.

There are a number of ways to denote a sequence, some of which are more useful for specific types of sequences. One way to specify a sequence is to list the elements. For example, the first four odd numbers form the sequence (1,3,5,7). This notation can be used for infinite sequences as well. For instance, the infinite sequence of positive odd integers can be written (1,3,5,7,...). Listing is most useful for infinite sequences with a pattern that can be easily discerned from the first few elements. Other ways to denote a sequence are discussed after the examples.

Important Examples

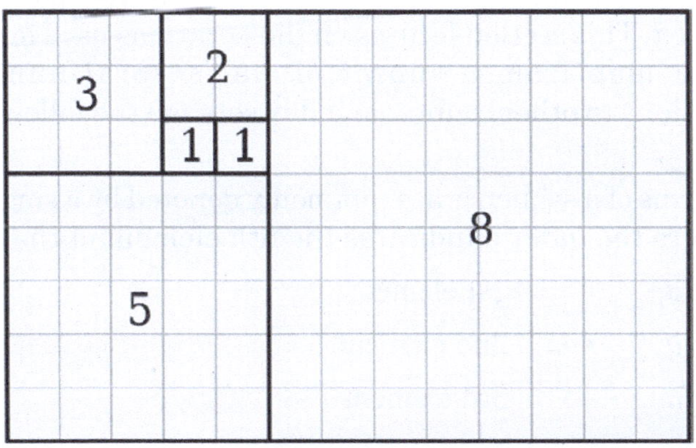

***Figure:** A tiling with squares whose sides are successive Fibonacci numbers in length.*

There are many important integer sequences. The prime numbers are the natural numbers bigger than 1, that have no divisors but 1 and themselves. Taking these in their natural order gives the sequence (2,3,5,7,11,13,17,...). The study of prime numbers has important applications for mathematics and specifically number theory.

The Fibonacci numbers are the integer sequence whose elements are the sum of the previous two elements. The first two elements are either 0 and 1 or 1 and 1 so that the sequence is (0,1,1,2,3,5,8,13,21,34,...).

Other interesting sequences include the ban numbers, whose spellings do not contain a certain letter of the alphabet. For instance, the eban numbers (do not contain 'e') form the sequence

(2,4,6,30,32,34,36,40,42,...). Another sequence based on the English spelling of the letters is the one based on their number of letters (3,3,5,4,4,3,5,5,4,3,6,6,8,...).

Other important examples of sequences include ones made up of rational numbers, real numbers, and complex numbers. The sequence (.9,.99,.999,.9999,...) approaches the number 1. In fact, every real number can be written as the limit of a sequence of rational numbers. For instance, for a sequence (3,3.1,3.14,3.141,3.1415,...) the limit of a sequence can be written as π. It is this fact that allows us to write any real number as the limit of a sequence of decimals. The decimal for π, however, does not have any pattern like the one for the sequence (0.9,0.99,...).

Indexing

Other notations can be useful for sequences whose pattern cannot be easily guessed, or for sequences that do not have a pattern such as the digits of π. This section focuses on the notations used for sequences that are a map from a subset of the natural numbers. For generalizations to other countable index sets see the following section and below.

The terms of a sequence are commonly denoted by a single variable, say a_n, where the *index* n indicates the nth element of the sequence.

$$
\begin{array}{lcl}
a_1 & \leftrightarrow & \text{1st element} \\
a_2 & \leftrightarrow & \text{2nd element} \\
a_3 & \leftrightarrow & \text{3rd element} \\
\vdots & & \vdots \\
a_{n-1} & \leftrightarrow & \text{(n-1)th element} \\
a_n & \leftrightarrow & \text{nth element} \\
a_{n+1} & \leftrightarrow & \text{(n+1)th element} \\
\vdots & & \vdots
\end{array}
$$

Indexing notation is used to refer to a sequence in the abstract. It is also a natural notation for sequences whose elements are related to the index n (the element's position) in a simple way. For instance, the sequence of the first 10 square numbers could be written as

$$(a_1, a_2, \ldots, a_{10}), \qquad a_k = k^2.$$

This represents the sequence (1,4,9,...100). This notation is often simplified further as

$$(a_k)_{k=1}^{10}, \qquad a_k = k^2.$$

Here the subscript {k=1} and superscript 10 together tell us that the elements of this sequence are the a_k such that $k = 1, 2, ..., 10$.

Sequences can be indexed beginning and ending from any integer. The infinity symbol ∞ is often used as the superscript to indicate the sequence including all integer k-values starting with a certain one. The sequence of all positive squares is then denoted

$$(a_k)_{k=1}^{\infty}, \qquad a_k = k^2.$$

In cases where the set of indexing numbers is understood, such as in analysis, the subscripts and superscripts are often left off. That is, one simply writes a_k for an arbitrary sequence. In analysis, k would be understood to run from 1 to ∞. However, sequences are often indexed starting from zero, as in

$$(a_k)_{k=0}^{\infty} = (a_0, a_1, a_2, ...).$$

In some cases the elements of the sequence are related naturally to a sequence of integers whose pattern can be easily inferred. In these cases the index set may be implied by a listing of the first few abstract elements. For instance, the sequence of squares of odd numbers could be denoted in any of the following ways.

- $(1, 9, 25, ...)$
- $(a_1, a_3, a_5, ...), \qquad a_k = k^2$
- $(a_{2k-1})_{k=1}^{\infty}, \qquad a_k = k^2$
- $(a_k)_{k=1}^{\infty}, \qquad a_k = (2k-1)^2$
- $((2k-1)^2)_{k=1}^{\infty}$

Moreover, the subscripts and superscripts could have been left off in the third, fourth, and fifth notations if the indexing set was understood to be the natural numbers.

Finally, sequences can more generally be denoted by writing a set inclusion in the subscript, such as in

$$(a_k)_{k \in \mathbf{N}}$$

The set of values that the index can take on is called the index set. In general, the ordering of the elements a_k is specified by the order of the elements in the indexing set. When N is the index set, the element a_{k+1} comes after the element a_k since in N, the element $(k+1)$ comes directly after the element k.

Specifying a Sequence by Recursion

Sequences whose elements are related to the previous elements in a straightforward way are often specified using recursion. This is in contrast to the specification of sequence elements in terms of their position.

To specify a sequence by recursion requires a rule to construct each consecutive element in terms of the ones before it. In addition, enough initial elements must be specified so that new elements of the sequence can be specified by the rule. The principle of mathematical induction can be used to prove that a sequence is well-defined, which is to say that every element of the sequence is specified at least once and has a single, unambiguous value. Induction can also be used to prove properties about a sequence, especially for sequences whose most natural specification is by recursion.

The Fibonacci sequence can be defined using a recursive rule along with two initial elements. The rule is that each element is the sum of the previous two elements, and the first two elements are 0 and 1.

$$a_n = a_{n-1} + a_{n-2}, \quad \text{with} \quad a_0 = 0 \text{ and } a_1 = 1.$$

The first ten terms of this sequence are 0,1,1,2,3,5,8,13,21, and 34. A more complicated example of a sequence that is defined recursively is Recaman's sequence, considered at the beginning of this section. We can define Recaman's sequence by

$a_0 = 0$ and $a_n = a_{n-1} - n$ if the result is positive and not already in the list. Otherwise, $a_n = a_{n-1} + n$.

Not all sequences can be specified by a rule in the form of an equation, recursive or not, and some can be quite complicated. For example, the sequence of prime numbers is the set of prime numbers in their natural order. This gives the sequence (2,3,5,7,11,13,17,...).

Definition and Basic Properties

There are many different notions of sequences in mathematics, some of which (*e.g.*, exact sequence) are not covered by the definitions and notations introduced below.

Definition

A sequence is usually defined as a function whose domain is a countable totally ordered set, although in many disciplines the domain is restricted, such as to the natural numbers. In real analysis a sequence is a function from a subset of the natural numbers to the real numbers.

In other words, a sequence is a map $f(n) : \mathrm{N} \rightarrow \mathrm{R}$. To recover our earlier notation we might identify $a_n = f(n)$ for all n or just write $a_n : \mathrm{N} \rightarrow \mathrm{R}$.

In complex analysis, sequences are defined as maps from the natural numbers to the complex numbers (C). In topology, sequences are often defined as functions from a subset of the natural numbers to a topological space. Sequences are an important concept for studying functions and, in topology, topological spaces. An important generalization of sequences, called a net, is to functions from a (possibly uncountable) directed set to a topological space.

Finite and Infinite

The length of a sequence is defined as the number of terms in the sequence.

A sequence of a finite length n is also called an n-tuple. Finite sequences include the empty sequence ( ) that has no elements.

Normally, the term *infinite sequence* refers to a sequence which is infinite in one direction, and finite in the other—the sequence has a first element, but no final element (a *singly infinite sequence*). A sequence that is infinite in both directions—it has neither a first nor a final element—is called a bi-infinite sequence, two-way infinite sequence, or doubly infinite sequence. For instance, a function from *all* integers into a set, such as the sequence of all even integers (..., –4, –2, 0, 2, 4, 6, 8...), is bi-infinite. This sequence could be denoted $(2n)_{n=-\infty}^{\infty}$. Formally, a bi-infinite sequence can be defined as a mapping from Z.

One can interpret singly infinite sequences as elements of the semigroup ring of the natural numbers R[N], and doubly infinite sequences as elements of the group ring of the integers R[Z]. This perspective is used in the Cauchy product of sequences.

Increasing and Decreasing

A sequence is said to be monotonically increasing if each term is greater than or equal to the one before it. For a sequence $(a_n)_{n=1}^{\infty}$ this can be written as $a_n \leq a_{n+1}$ for all $n \in \mathrm{N}$. If each consecutive term is strictly greater than (>) the previous term then the sequence is called strictly monotonically increasing. A sequence is monotonically decreasing if each consecutive term is less than or equal to the previous one, and strictly monotonically decreasing if each is strictly less than the previous. If a sequence is either increasing or decreasing it is called a monotone sequence. This is a special case of the more general notion of a monotonic function.

The terms nondecreasing and nonincreasing are often used in place of *increasing* and *decreasing* in order to avoid any possible confusion with *strictly increasing* and *strictly decreasing*, respectively.

Bounded

If the sequence of real numbers (a_n) is such that all the terms, after a certain one, are less than some real number M, then the sequence is said to be bounded from above. In less words, this means $a_n \leq M$ for all n greater than N for some pair M and N. Any such M is called an *upper bound.*

Likewise, if, for some real m, a_n e– m for all n greater than some N, then the sequence is bounded from below and any such m is called a *lower bound.* If a sequence is both bounded from above and bounded from below then the sequence is said to be bounded.

Other Types of Sequences

A subsequence of a given sequence is a sequence formed from the given sequence by deleting some of the elements without disturbing the relative positions of the remaining elements. For instance, the sequence of positive even integers (2,4,6,...) is a subsequence of the positive integers (1,2,3,...). The positions of some elements change when other elements are deleted. However, the relative positions are preserved.

Some other types of sequences that are easy to define include:

- An integer sequence is a sequence whose terms are integers.
- A polynomial sequence is a sequence whose terms are polynomials.
- A positive integer sequence is sometimes called multiplicative if $a_{nm} = a_n a_m$ for all pairs n,m such that n and m are coprime. In other instances, sequences are often called *multiplicative* if $a_n = na_1$ for all n. Moreover, the multiplicative Fibonacci sequence satisfies the recursion relation $a_n = a_{n-1} a_{n-2}$.

Limits and Convergence

One of the most important properties of a sequence is *convergence.* Informally, a sequence converges if it has a *limit.* Continuing informally, a (singly infinite) sequence has a limit if it approaches some value L, called the limit, as n becomes very large.

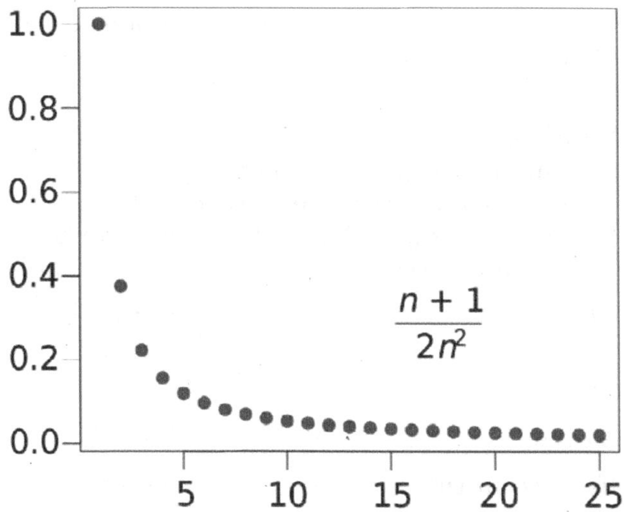

Figure: *The plot of a convergent sequence* (a_n) *is shown in blue. Visually we can see that the sequence is converging to the limit zero as* n *increases.*

That is, for an abstract sequence (a_n) (with n running from 1 to infinity understood) the value of a_n approaches L as $n \to \infty$, denoted

$$\lim_{n\to\infty} a_n = L.$$

More precisely, the sequence converges if there exists a limit L such that the remaining a_n's are arbitrarily close to L for some n large enough.

If a sequence converges to some limit, then it is convergent; otherwise it is divergent.

If a_n gets arbitrarily large as $n \to \infty$ we write

$$\lim_{n\to\infty} a_n = \infty.$$

In this case the sequence (a_n) *diverges*, or that it converges to infinity.

If a_n becomes arbitrarily "small" negative numbers (large in magnitude) as $n \to \infty$ we write

$$\lim_{n\to\infty} a_n = -\infty$$

and say that the sequence diverges or converges to minus infinity.

Definition of Convergence

For sequences that can be written as $(a_n)_{n=1}^{\infty}$ with $a_n \in$ R we can write (a_n) with the indexing set understood as N. These sequences are most common in real analysis. The generalizations to other types of sequences are considered in the following section and the main page Limit of a sequence.

Let (a_n) be a sequence. In words, the sequence (a_n) is said to *converge* if there exists a number L such that no matter how close we want the a_n to be to L (say ε-close where $\varepsilon > 0$), we can find a natural number N such that all terms $(a_{N+1}, a_{N+2}, \ldots)$ are further closer to L (within ε of L). This is often written more compactly using symbols. For instance,

for all $\varepsilon > 0$, there exists a natural number N such that $L-\varepsilon < a_n <$ L+ε for all n e– N.

In even more compact notation

$$\forall \epsilon > 0, \exists N \in \mathbf{N} \text{ s.t. } \forall n \geq N, |a_n - L| < \epsilon.$$

The difference in the definitions of convergence for (one-sided) sequences in complex analysis and metric spaces is that the absolute value $|a_n - L|$ is interpreted as the distance in the complex plane ($\sqrt{z^* z}$), and the distance under the appropriate metric, respectively.

Applications and Important Results

Important results for convergence and limits of (one-sided) sequences of real numbers include the following. These equalities are all true at least when both sides exist. For a discussion of when the existence of the limit on one side implies the existence of the other see a real analysis text such as can be found in the references.

- The limit of a sequence is unique.
- $\lim\limits_{n\to\infty}(a_n \pm b_n) = \lim\limits_{n\to\infty} a_n \pm \lim\limits_{n\to\infty} b_n$
- $\lim\limits_{n\to\infty} ca_n = c \lim\limits_{n\to\infty} a_n$
- $\lim\limits_{n\to\infty}(a_n b_n) = (\lim\limits_{n\to\infty} a_n)(\lim\limits_{n\to\infty} b_n)$
- $\lim\limits_{n\to\infty} \dfrac{a_n}{b_n} = \dfrac{\lim\limits_{n\to\infty} a_n}{\lim\limits_{n\to\infty} b_n}$ provided $\lim\limits_{n\to\infty} b_n \neq 0$
- $\lim\limits_{n\to\infty} a_n^p = \left[\lim\limits_{n\to\infty} a_n\right]^p$
- If $a_n \leq b_n$ for all n greater than some N, then $\lim\limits_{n\to\infty} a_n \leq \lim\limits_{n\to\infty} b_n$.
- (Squeeze Theorem) If $a_n \leq c_n \leq b_n$ for all $n > N$, and $\lim\limits_{n\to\infty} a_n = \lim\limits_{n\to\infty} b_n = L$, then $\lim\limits_{n\to\infty} c_n = L$.

- If a sequence is bounded and monotonic then it is convergent.
- A sequence is convergent if and only if every subsequence is convergent.

Cauchy Sequences

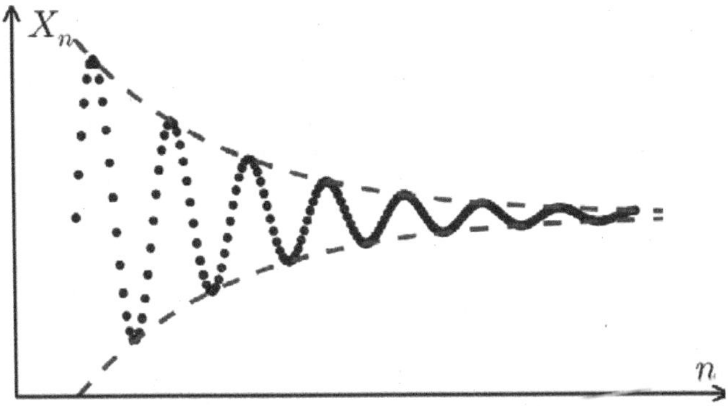

The plot of a Cauchy sequence (X_n), shown in blue, as X_n versus n. Visually, we see that the sequence appears to be converging to the limit zero as the terms in the sequence become closer together as n increases. In the real numbers every Cauchy sequence converges to some limit.

A Cauchy sequence is a sequence whose terms become arbitrarily close together as n gets very large. The notion of a Cauchy sequence is important in the study of sequences in metric spaces, and, in particular, in real analysis. One particularly important result in real analysis is *Cauchy characterization of convergence for sequences*:

In the real numbers, a sequence is convergent if and only if it is Cauchy.

In contrast, in the rational numbers, e.g. the sequence defined by $x_1 = 1$ and $x_{n+1} = x_n + 2/x_n/2$ is Cauchy, but has no rational limit, cf. here.

Series

A *series* is, informally speaking, the sum of the terms of a sequence. That is, adding the first N terms of a (one-sided) sequence forms the Nth term of another sequence, called a *series*. Thus the N series of the sequence (a_n) results in another sequence (S_N) given by:

$$\begin{aligned}
S_1 &= a_1 \\
S_2 &= a_1 + a_2 \\
S_3 &= a_1 + a_2 + a_3 \\
\vdots &\quad \vdots \\
S_N &= a_1 + a_2 + a_3 + \cdots \\
\vdots &\quad \vdots
\end{aligned}$$

We can also write the nth term of the series as

$$S_N = \sum_{n=1}^{N} a_n.$$

Then the concepts used to talk about sequences, such as convergence, carry over to series (the sequence of partial sums) and the properties can be characterized as properties of the underlying sequences (such as (a_n) in the last example). The limit, if it exists, of an infinite series (the series created from an infinite sequence) is written as

$$\lim_{N\to\infty} S_N = \sum_{n=1}^{\infty} a_n.$$

Use in Other Fields of Mathematics

This section is incomplete. *(January 2013)*

Vectors

Sequences over a field may also be viewed as vectors in a vector space. Specifically, the set of F-valued sequences (where F is a field) is a function space (in fact, a product space) of F-valued functions over the set of natural numbers.

Ordinal-Indexed Sequence

An ordinal-indexed sequence is a generalization of a sequence. If α is a limit ordinal and X is a set, an α-indexed sequence of elements of X is a function from α to X. In this terminology an ω-indexed sequence is an ordinary sequence.

Sequences and Automata

Automata or finite state machines can typically be thought of as directed graphs, with edges labelled using some specific alphabet, Σ. Most familiar types of automata transition from state to state by reading input letters from Σ, following edges with matching labels; the ordered input for such an automaton forms a sequence called a *word* (or input word). The sequence of states encountered by the automaton when processing a word is called a *run*.

A nondeterministic automaton may have unlabelled or duplicate out-edges for any state, giving more than one successor for some input letter. This is typically thought of as producing multiple possible runs for a given word, each being a sequence of single states, rather than producing a single run that is a sequence of sets of states; however, 'run' is occasionally used to mean the latter.

Analysis

In analysis, when talking about sequences, one will generally consider sequences of the form

$$(x_1, x_2, x_3, \ldots) \text{ or } (x_0, x_1, x_2, \ldots)$$

which is to say, infinite sequences of elements indexed by natural numbers.

It may be convenient to have the sequence start with an index different from 1 or 0. For example, the sequence defined by $x_n = 1/\log(n)$ would be defined only for n e– 2. When talking about such infinite sequences, it is usually sufficient (and does not change much for most considerations) to assume that the members of the sequence are defined at least for all indices large enough, that is, greater than some given N.

The most elementary type of sequences are numerical ones, that is, sequences of real or complex numbers. This type can be generalized to sequences of elements of some vector space. In analysis, the vector spaces considered are often function spaces. Even more generally, one can study sequences with elements in some topological space.

Infinite Sequences in Theoretical Computer Science

Infinite sequences of digits (or characters) drawn from a finite alphabet are of particular interest in theoretical computer science. They are often referred to simply as *sequences* or *streams*, as opposed to finite strings. Infinite binary sequences, for instance, are infinite sequences of bits (characters drawn from the alphabet {0, 1}). The set $C = \{0, 1\}^\infty$ of all infinite, binary sequences is sometimes called the Cantor space.

An infinite binary sequence can represent a formal language (a set of strings) by setting the n th bit of the sequence to 1 if and only if the nth string (in shortlex order) is in the language. Therefore, the study of complexity classes, which are sets of languages, may be regarded as studying sets of infinite sequences.

An infinite sequence drawn from the alphabet $\{0, 1, \ldots, b - 1\}$ may also represent a real number expressed in the base-b positional number system. This equivalence is often used to bring the techniques of real analysis to bear on complexity classes. In particular, the term sequence space usually refers to a linear subspace of the set of all possible infinite sequences with elements in C.

Abstract Algebra

If A is a set, the free monoid over A (denoted A^*, also called Kleene star of A) is a monoid containing all the finite sequences (or strings) of zero or more elements of A, with the binary operation of concatenation. The free semigroup A^+ is the subsemigroup of A^* containing all elements except the empty sequence.

Types

- ±1-sequence
- Arithmetic progression
- Cauchy sequence
- Farey sequence
- Fibonacci sequence
- Geometric progression
- Look-and-say sequence
- Thue–Morse sequence

Related Concepts

- List (computing)
- Ordinal-indexed sequence
- Recursion (computer science)
- Tuple
- Set theory

Operations

- Cauchy product
- Limit of a sequence

Series

A series is, informally speaking, the sum of the terms of a sequence. Finite sequences and series have defined first and last terms, whereas infinite sequences and series continue indefinitely.

In mathematics, given an infinite sequence of numbers $\{ a_n \}$, a series is informally the result of adding all those terms together: $a_1 + a_2 + a_3 + \cdots$. These can be written more compactly using the summation symbol $\sum$. An example is the famous series from Zeno's dichotomy and its mathematical representation:

$$\sum_{n=1}^{\infty} \frac{1}{2^n} = \frac{1}{2} + \frac{1}{4} + \frac{1}{8} + \cdots.$$

The terms of the series are often produced according to a certain rule, such as by a formula, or by an algorithm. As there are an infinite number of terms, this notion is often called an infinite series. Unlike finite summations, infinite series need tools from mathematical analysis, and specifically the notion of limits, to be fully understood and manipulated. In addition to their ubiquity in mathematics, infinite series are also widely used in other quantitative disciplines such as physics, computer science, and finance.

Definition

For any sequence $\{a_n\}$ of rational numbers, real numbers, complex numbers, functions thereof, etc., the associated series is defined as the ordered formal sum

$$\sum_{n=0}^{\infty} a_n = a_0 + a_1 + a_2 + \cdots .$$

The sequence of partial sums $\{S_k\}$ associated to a series $\sum_{n=0}^{\infty} a_n$ is defined for each k as the sum of the sequence $\{a_n\}$ from a_0 to a_k

$$S_k = \sum_{n=0}^{k} a_n = a_0 + a_1 + \cdots + a_k .$$

By definition the series $\sum_{n=0}^{\infty} a_n$ converges to a limit L if and only if the associated sequence of partial sums $\{S_k\}$ converges to L. This definition is usually written as

$$L = \sum_{n=0}^{\infty} a_n \Leftrightarrow L = \lim_{k \to \infty} S_k .$$

More generally, if $I \xrightarrow{a} G$ is a function from an index set I to a set G, then the series associated to a is the formal sum of the elements $a(x) \in G$ over the index elements $x \in I$ denoted by the

$$\sum_{x \in I} a(x).$$

When the index set is the natural numbers $I = \mathbb{N}$, the function $\mathbb{N} \xrightarrow{a} G$ is a sequence denoted by $a(n) = a_n$. A series indexed on the natural numbers is an ordered formal sum and so we rewrite as $\sum_{n=0}^{\infty}$ in order to emphasize the ordering induced by the natural numbers. Thus, we obtain the common notation for a series indexed by the natural numbers

$$\sum_{n=0}^{\infty} a_n = a_0 + a_1 + a_2 + \cdots.$$

When the set G is a semigroup, the sequence of partial sums $\{S_k\} \subset G$ associated to a sequence $\{a_n\} \subset G$ is defined for each k as the sum of the terms

$$S_k = \sum_{n=0}^{k} a_n = a_0 + a_1 + \cdots + a_k.$$

When the semigroup G is also a topological space, then the series $\sum_{n=0}^{\infty} a_n$ converges to an element $L \in G$ if and only if the associated sequence of partial sums $\{S_k\}$ converges to . This definition is usually written as

$$L = \sum_{n=0}^{\infty} a_n \Leftrightarrow L = \lim_{k\to\infty} S_k.$$

Convergent Series

A series $\sum a_n$ is said to 'converge' or to 'be convergent' when the sequence S_N of partial sums has a finite limit. If the limit of S_N is infinite or does not exist, the series is said to diverge. When the limit of partial sums exists, it is called the sum of the series

$$\sum_{n=0}^{\infty} a_n = \lim_{N\to\infty} S_N = \lim_{N\to\infty} \sum_{n=0}^{N} a_n.$$

An easy way that an infinite series can converge is if all the a_n are zero for n sufficiently large. Such a series can be identified with a finite sum, so it is only infinite in a trivial sense.

Working out the properties of the series that converge even if infinitely many terms are non-zero is the essence of the study of series. Consider the example

$$1 + \frac{1}{2} + \frac{1}{4} + \frac{1}{8} + \cdots + \frac{1}{2^n} + \cdots.$$

It is possible to "visualize" its convergence on the real number line: we can imagine a line of length 2, with successive segments marked off of lengths 1, ½, ¼, etc. There is always room to mark the next segment, because the amount of line remaining is always the same as the last segment marked: when we have marked off ½, we still have a piece of length ½ unmarked, so we can certainly mark the next ¼.

This argument does not prove that the sum is *equal* to 2 (although it is), but it does prove that it is *at most* 2. In other words, the series has an upper bound. Given that the series converges, proving that it is equal to 2 requires only elementary algebra. If the series is denoted S, it can be seen that

$$S/2 = \frac{1+\frac{1}{2}+\frac{1}{4}+\frac{1}{8}+\cdots}{2} = \frac{1}{2}+\frac{1}{4}+\frac{1}{8}+\frac{1}{16}+\cdots.$$

Therefore,

$$S - S/2 = 1 \Rightarrow S = 2.$$

Mathematicians extend the idiom discussed earlier to other, equivalent notions of series. For instance, when we talk about a recurring decimal, as in

$$x = 0.111\ldots$$

we are talking, in fact, just about the series

$$\sum_{n=1}^{\infty}\frac{1}{10^n}.$$

But since these series always converge to real numbers (because of what is called the completeness property of the real numbers), to talk about the series in this way is the same as to talk about the numbers for which they stand. In particular, it should offend no sensibilities if we make no distinction between 0.111... and $^1/_9$. Less clear is the argument that 9 × 0.111... = 0.999... = 1, but it is not untenable when we consider that we can formalize the proof knowing only that limit laws preserve the arithmetic operations.

Examples

- A geometric series is one where each successive term is produced by multiplying the previous term by a constant number (called the common ratio in this context). Example:

  $$1+\frac{1}{2}+\frac{1}{4}+\frac{1}{8}+\frac{1}{16}+\cdots = \sum_{n=0}^{\infty}\frac{1}{2^n}.$$

  In general, the geometric series

  $$\sum_{n=0}^{\infty} z^n$$

  converges if and only if $|z| < 1$.

- An Arithmetico-geometric sequence is a generalization of the geometric series, which has coefficients of the common ratio equal to the terms in an arithmetic series. Example:

$$3+\frac{5}{2}+\frac{7}{4}+\frac{9}{8}+\frac{11}{16}+\cdots=\sum_{n=0}^{\infty}\frac{(3+2n)}{2^n}.$$

- The harmonic series is the series

$$1+\frac{1}{2}+\frac{1}{3}+\frac{1}{4}+\frac{1}{5}+\cdots=\sum_{n=1}^{\infty}\frac{1}{n}.$$

The harmonic series is divergent.

- An alternating series is a series where terms alternate signs. Example:

$$1-\frac{1}{2}+\frac{1}{3}-\frac{1}{4}+\frac{1}{5}-\cdots=\sum_{n=1}^{\infty}(-1)^{n+1}\frac{1}{n}=\ln(2).$$

- The p-series

$$\sum_{n=1}^{\infty}\frac{1}{n^r}$$

converges if $r > 1$ and diverges for $r \leq 1$, which can be shown with the integral criterion described below in convergence tests. As a function of r, the sum of this series is Riemann's zeta function.

- A telescoping series

$$\sum_{n=1}^{\infty}(b_n - b_{n+1})$$

converges if the sequence b_n converges to a limit L as n goes to infinity. The value of the series is then $b_1 - L$.

Calculus and Partial Summation as an Operation on Sequences

Partial summation takes as input a sequence, $\{ a_n \}$, and gives as output another sequence, $\{ S_N \}$. It is thus a unary operation on sequences. Further, this function is linear, and thus is a linear operator on the vector space of sequences, denoted Σ. The inverse operator is the finite difference operator, Δ. These behave as discrete analogs of integration and differentiation, only for series (functions of a natural number) instead of functions of a real variable. For example, the sequence {1, 1, 1, ...} has series {1, 2, 3, 4, ...} as its partial summation, which is analogous to the fact that $\int_0^x 1dt = x$.

In computer science it is known as prefix sum.

Properties of Series

Series are classified not only by whether they converge or diverge, but also by the properties of the terms a_n (absolute or conditional convergence); type of convergence of the series (pointwise, uniform); the class of the term a_n (whether it is a real number, arithmetic progression, trigonometric function); etc.

Non-Negative Terms

When a_n is a non-negative real number for every n, the sequence S_N of partial sums is non-decreasing. It follows that a series $\sum a_n$ with non-negative terms converges if and only if the sequence S_N of partial sums is bounded.

For example, the series

$$\sum_{n\geq 1} \frac{1}{n^2}$$

is convergent, because the inequality

$$\frac{1}{n^2} \leq \frac{1}{n-1} - \frac{1}{n}, \quad n \geq 2,$$

and a telescopic sum argument implies that the partial sums are bounded by 2.

Absolute Convergence

A series

$$\sum_{n=0}^{\infty} a_n$$

is said to converge absolutely if the series of absolute values

$$\sum_{n=0}^{\infty} |a_n|$$

converges. This is sufficient to guarantee not only that the original series converges to a limit, but also that any reordering of it converges to the same limit.

Conditional Convergence

A series of real or complex numbers is said to be conditionally convergent (or semi-convergent) if it is convergent but not absolutely convergent. A famous example is the alternating series

$$\sum_{n=1}^{\infty} \frac{(-1)^{n+1}}{n} = 1 - \frac{1}{2} + \frac{1}{3} - \frac{1}{4} + \frac{1}{5} - \cdots$$

which is convergent (and its sum is equal to ln 2), but the series formed by taking the absolute value of each term is the divergent harmonic series. The Riemann series theorem says that any conditionally convergent series can be reordered to make a divergent series, and moreover, if the a_n are real and S is any real number, that one can find a reordering so that the reordered series converges with sum equal to S.

Abel's test is an important tool for handling semi-convergent series. If a series has the form

$$\sum a_n = \sum \lambda_n b_n$$

where the partial sums $B_N = b_0 + \cdots + b_n$ are bounded, λ_n has bounded variation, and $\lim \lambda_n B_n$ exists:

$$\sup_N \left|\sum_{n=0}^{N} b_n\right| < \infty, \ \sum |\lambda_{n+1} - \lambda_n| < \infty \text{ and } \lambda_n B_n \text{ converges,}$$

then the series $\sum a_n$ is convergent. This applies to the pointwise convergence of many trigonometric series, as in

$$\sum_{n=2}^{\infty} \frac{\sin(nx)}{\ln n}$$

with $0 < x < 2\pi$. Abel's method consists in writing $b_{n+1} = B_{n+1} - B_n$, and in performing a transformation similar to integration by parts (called summation by parts), that relates the given series – a_n to the absolutely convergent series

$$\sum (\lambda_n - \lambda_{n+1}) B_n.$$

Convergence Tests

- n-th term test: If $\lim_{n \to -} a_n \neq 0$ then the series diverges.
- Comparison test 1: If $-b_n$ is an absolutely convergent series such that $|a_n| \leq C |b_n|$ for some number C and for sufficiently large n, then $-a_n$ converges absolutely as well. If $-|b_n|$ diverges, and $|a_n| \leq |b_n|$ for all sufficiently large n, then $-a_n$ also fails to converge absolutely (though it could still be conditionally convergent, e.g. if the a_n alternate in sign).
- Comparison test 2: If $-b_n$ is an absolutely convergent series such that $|a_{n+1}/a_n| \leq |b_{n+1}/b_n|$ for sufficiently large n, then $-a_n$ converges absolutely as well. If $-|b_n|$ diverges, and $|a_{n+1}/a_n| \leq |b_{n+1}/b_n|$ for all sufficiently large n, then $-a_n$ also fails to converge absolutely (though it could still be conditionally convergent, e.g. if the a_n alternate in sign).

- Ratio test: If there exists a constant $C < 1$ such that $|a_{n+1}/a_n|<C$ for all sufficiently large n, then $-a_n$ converges absolutely. When the ratio is less than 1, but not less than a constant less than 1, convergence is possible but this test does not establish it.
- Root test: If there exists a constant $C < 1$ such that $|a_n|^{1/n} \leq C$ for all sufficiently large n, then $-a_n$ converges absolutely.
- Integral test: if $f(x)$ is a positive monotone decreasing function defined on the interval [1,) with $f(n) = a_n$ for all n, then $-a_n$ converges if and only if the integral $+-_1^- f(x)\, dx$ is finite.
- Cauchy's condensation test: If a_n is non-negative and non-increasing, then the two series $-a_n$ and $-2^k a_{(2^k)}$ are of the same nature: both convergent, or both divergent.
- Alternating series test: A series of the form $\sum (-1)^n a_n$ (with $a_n \geq 0$) is called *alternating*. Such a series converges if the sequence a_n is monotone decreasing and converges to 0. The converse is in general not true.
- For some specific types of series there are more specialised convergence tests, for instance for Fourier series there is the Dini test.

Series of Functions

A series of real- or complex-valued functions

$$\sum_{n=0}^{\infty} f_n(x)$$

converges pointwise on a set E, if the series converges for each x in E as an ordinary series of real or complex numbers. Equivalently, the partial sums

$$s_N(x) = \sum_{n=0}^{N} f_n(x)$$

converge to $f(x)$ as $N \to \infty$ for each $x \in E$.

A stronger notion of convergence of a series of functions is called uniform convergence. The series converges uniformly if it converges pointwise to the function $f(x)$, and the error in approximating the limit by the Nth partial sum,

$$|s_N(x) - f(x)|$$

can be made minimal *independently* of x by choosing a sufficiently large N.

Uniform convergence is desirable for a series because many properties of the terms of the series are then retained by the limit. For example, if a series of continuous functions converges uniformly, then the limit function is also continuous. Similarly, if the f_n are integrable on a closed and bounded interval I and converge uniformly, then the series is also integrable on I and can be integrated term-by-term. Tests for uniform convergence include the Weierstrass' M-test, Abel's uniform convergence test, Dini's test.

More sophisticated types of convergence of a series of functions can also be defined. In measure theory, for instance, a series of functions converges almost everywhere if it converges pointwise except on a certain set of measure zero. Other modes of convergence depend on a different metric space structure on the space of functions under consideration. For instance, a series of functions converges in mean on a set E to a limit function f provided

$$\int_E |s_N(x) - f(x)|^2 \, dx \to 0$$

as $N \to \infty$..

Power Series

A power series is a series of the form

$$\sum_{n=0}^{\infty} a_n (x-c)^n.$$

The Taylor series at a point c of a function is a power series that, in many cases, converges to the function in a neighbourhood of c. For example, the series

$$\sum_{n=0}^{\infty} \frac{x^n}{n!}$$

is the Taylor series of e^x at the origin and converges to it for every x.

Unless it converges only at $x=c$, such a series converges on a certain open disc of convergence centered at the point c in the complex plane, and may also converge at some of the points of the boundary of the disc. The radius of this disc is known as the radius of convergence, and can in principle be determined from the asymptotics of the coefficients a_n. The convergence is uniform on closed and bounded (that is, compact) subsets of the interior of the disc of convergence: to wit, it is uniformly convergent on compact sets.

Historically, mathematicians such as Leonhard Euler operated liberally with infinite series, even if they were not convergent. When calculus was put on a sound and correct foundation in the nineteenth century, rigorous proofs of the convergence of series were always required. However, the formal operation with non-convergent series has been retained in rings of formal power series which are studied in abstract algebra.

Formal power series are also used in combinatorics to describe and study sequences that are otherwise difficult to handle; this is the method of generating functions.

Laurent Series

Laurent series generalize power series by admitting terms into the series with negative as well as positive exponents. A Laurent series is thus any series of the form

$$\sum_{n=-\infty}^{\infty} a_n x^n.$$

If such a series converges, then in general it does so in an annulus rather than a disc, and possibly some boundary points. The series converges uniformly on compact subsets of the interior of the annulus of convergence.

Dirichlet Series

A Dirichlet series is one of the form

$$\sum_{n=1}^{\infty} \frac{a_n}{n^s},$$

where s is a complex number. For example, if all a_n are equal to 1, then the Dirichlet series is the Riemann zeta function

$$\zeta(s) = \sum_{n=1}^{\infty} \frac{1}{n^s}.$$

Like the zeta function, Dirichlet series in general play an important role in analytic number theory. Generally a Dirichlet series converges if the real part of s is greater than a number called the abscissa of convergence. In many cases, a Dirichlet series can be extended to an analytic function outside the domain of convergence by analytic continuation. For example, the Dirichlet series for the zeta function converges absolutely when Re $s > 1$, but the zeta function can be extended to a holomorphic function defined on $\mathbf{C} \setminus \{1\}$ with a simple pole at 1.

This series can be directly generalized to general Dirichlet series.

Trigonometric Series

A series of functions in which the terms are trigonometric functions is called a trigonometric series:

$$\tfrac{1}{2}A_0 + \sum_{n=1}^{\infty}(A_n \cos nx + B_n \sin nx).$$

The most important example of a trigonometric series is the Fourier series of a function.

Development of Infinite Series

Greek mathematician Archimedes produced the first known summation of an infinite series with a method that is still used in the area of calculus today. He used the method of exhaustion to calculate the area under the arc of a parabola with the summation of an infinite series, and gave a remarkably accurate approximation of π.

In the 17th century, James Gregory worked in the new decimal system on infinite series and published several Maclaurin series. In 1715, a general method for constructing the Taylor series for all functions for which they exist was provided by Brook Taylor. Leonhard Euler in the 18th century, developed the theory of hypergeometric series and q-series.

Convergence Criteria

The investigation of the validity of infinite series is considered to begin with Gauss in the 19th century. Euler had already considered the hypergeometric series

$$1+\frac{\alpha\beta}{1\cdot\gamma}x+\frac{\alpha(\alpha+1)\beta(\beta+1)}{1\cdot 2\cdot\gamma(\gamma+1)}x^2+\cdots$$

on which Gauss published a memoir in 1812. It established simpler criteria of convergence, and the questions of remainders and the range of convergence.

Cauchy (1821) insisted on strict tests of convergence; he showed that if two series are convergent their product is not necessarily so, and with him begins the discovery of effective criteria. The terms *convergence* and *divergence* had been introduced long before by Gregory (1668). Leonhard Euler and Gauss had given various criteria, and Colin Maclaurin had anticipated some of Cauchy's discoveries. Cauchy advanced the theory of power series by his expansion of a complex function in such a form.

Abel (1826) in his memoir on the binomial series

$$1+\frac{m}{1!}x+\frac{m(m-1)}{2!}x^2+\cdots$$

corrected certain of Cauchy's conclusions, and gave a completely scientific summation of the series for complex values of m and x. He showed the necessity of considering the subject of continuity in questions of convergence.

Cauchy's methods led to special rather than general criteria, and the same may be said of Raabe (1832), who made the first elaborate investigation of the subject, of De Morgan (from 1842), whose logarithmic test DuBois-Reymond (1873) and Pringsheim (1889) have shown to fail within a certain region; of Bertrand (1842), Bonnet (1843), Malmsten (1846, 1847, the latter without integration); Stokes (1847), Paucker (1852), Chebyshev (1852), and Arndt (1853).

General criteria began with Kummer (1835), and have been studied by Eisenstein (1847), Weierstrass in his various contributions to the theory of functions, Dini (1867), DuBois-Reymond (1873), and many others. Pringsheim's memoirs (1889) present the most complete general theory.

Uniform Convergence

The theory of uniform convergence was treated by Cauchy (1821), his limitations being pointed out by Abel, but the first to attack it successfully were Seidel and Stokes (1847–48). Cauchy took up the problem again (1853), acknowledging Abel's criticism, and reaching the same conclusions which Stokes had already found. Thomae used the doctrine (1866), but there was great delay in recognising the importance of distinguishing between uniform and non-uniform convergence, in spite of the demands of the theory of functions.

Semi-Convergence

A series is said to be semi-convergent (or conditionally convergent) if it is convergent but not absolutely convergent.

Semi-convergent series were studied by Poisson (1823), who also gave a general form for the remainder of the Maclaurin formula. The most important solution of the problem is due, however, to Jacobi (1834), who attacked the question of the remainder from a different standpoint and reached a different formula. This expression was also worked out, and another one given, by Malmsten (1847). Schlφmilch (*Zeitschrift*, Vol.I, p. 192, 1856) also improved Jacobi's remainder, and showed the relation between the remainder and Bernoulli's function

$$F(x) = 1^n + 2^n + \cdots + (x-1)^n.$$

Genocchi (1852) has further contributed to the theory.

Among the early writers was Wronski, whose "loi suprême" (1815) was hardly recognised until Cayley (1873) brought it into prominence.

Fourier Series

Fourier series were being investigated as the result of physical considerations at the same time that Gauss, Abel, and Cauchy were working out the theory of infinite series. Series for the expansion of sines and cosines, of multiple arcs in powers of the sine and cosine of the arc had been treated by Jacob Bernoulli (1702) and his brother Johann Bernoulli (1701) and still earlier by Vieta. Euler and Lagrange simplified the subject, as did Poinsot, Schröter, Glaisher, and Kummer.

Fourier (1807) set for himself a different problem, to expand a given function of x in terms of the sines or cosines of multiples of x, a problem which he embodied in his Théorie analytique de la chaleur (1822). Euler had already given the formulas for determining the coefficients in the series; Fourier was the first to assert and attempt to prove the general theorem.

Poisson (1820–23) also attacked the problem from a different standpoint. Fourier did not, however, settle the question of convergence of his series, a matter left for Cauchy (1826) to attempt and for Dirichlet (1829) to handle in a thoroughly scientific manner. Dirichlet's treatment (Crelle, 1829), of trigonometric series was the subject of criticism and improvement by Riemann (1854), Heine, Lipschitz, Schläfli, and du Bois-Reymond. Among other prominent contributors to the theory of trigonometric and Fourier series were Dini, Hermite, Halphen, Krause, Byerly and Appell.

Generalizations

Asymptotic Series: Asymptotic series, otherwise asymptotic expansions, are infinite series whose partial sums become good approximations in the limit of some point of the domain. In general they do not converge. But they are useful as sequences of approximations, each of which provides a value close to the desired answer for a finite number of terms. The difference is that an asymptotic series cannot be made to produce an answer as exact as desired, the way that convergent series can. In fact, after a certain number of terms, a typical asymptotic series reaches its best approximation; if more terms are included, most such series will produce worse answers.

Divergent Series

Under many circumstances, it is desirable to assign a limit to a series which fails to converge in the usual sense. A summability method is such an assignment of a limit to a subset of the set of divergent series which properly extends the classical notion of convergence. Summability methods include Cesàro summation, (C,k) summation, Abel summation, and Borel summation, in increasing order of generality (and hence applicable to increasingly divergent series).

A variety of general results concerning possible summability methods are known. The Silverman–Toeplitz theorem characterizes *matrix summability methods*, which are methods for summing a divergent series by applying an infinite matrix to the vector of coefficients. The most general method for summing a divergent series is non-constructive, and concerns Banach limits.

Series in Banach Spaces

The notion of series can be easily extended to the case of a Banach space. If x_n is a sequence of elements of a Banach space X, then the series Σx_n converges to $x \in X$ if the sequence of partial sums of the series tends to x; to wit,

$$\| x - \sum_{n=0}^{N} x_n \| \to 0$$

as $N \to \infty$.

More generally, convergence of series can be defined in any abelian Hausdorff topological group. Specifically, in this case, Σx_n converges to x if the sequence of partial sums converges to x.

Summations over Arbitrary Index Sets

Definitions may be given for sums over an arbitrary index set I. There are two main differences with the usual notion of series: first, there is no specific order given on the set I; second, this set I may be uncountable.

Families of Non-Negative Numbers

When summing a family $\{a_i\}$, $i \in I$, of non-negative numbers, one may define

$$\sum_{i \in I} a_i = \sup\left\{\sum_{i \in A} a_i \,\middle|\, A \text{ finite}, A \subset I\right\} \in [0, +\infty].$$

When the sum is finite, the set of $i \in I$ such that $a_i > 0$ is countable. Indeed for every $n \geq 1$, the set $A_n = \{i \in I : a_i > 1/n\}$ is finite, because

$$\frac{1}{n}\text{card}(A_n) \le \sum_{i\in A_n} a_i \le \sum_{i\in I} a_i < \infty.$$

If I is countably infinite and enumerated as $I = \{i_0, i_1, ...\}$ then the above defined sum satisfies

$$\sum_{i\in I} a_i = \sum_{k=0}^{+\infty} a_{i_k},$$

provided the value ∞ is allowed for the sum of the series.

Any sum over non-negative reals can be understood as the integral of a non-negative function with respect to the counting measure, which accounts for the many similarities between the two constructions.

Abelian Topological Groups

Let $a : I \to X$, where I is any set and X is an abelian Hausdorff topological group. Let F be the collection of all finite subsets of I. Note that F is a directed set ordered under inclusion with union as join. Define the sum S of the family a as the limit

$$S = \sum_{i\in I} a_i = \lim\left\{\sum_{i\in A} a_i \middle| A \in F\right\}$$

if it exists and say that the family a is unconditionally summable. Saying that the sum S is the limit of finite partial sums means that for every neighbourhood V of 0 in X, there is a finite subset A_0 of I such that

$$S - \sum_{i\in A} a_i \in V, \quad A \supset A_0.$$

Because F is not totally ordered, this is not a limit of a sequence of partial sums, but rather of a net.

For every W, neighbourhood of 0 in X, there is a smaller neighbourhood V such that $V - V \subset W$. It follows that the finite partial sums of an unconditionally summable family a_i, $i \in I$, form a *Cauchy net*, that is: for every W, neighbourhood of 0 in X, there is a finite subset A_0 of I such that

$$\sum_{i\in A_1} a_i - \sum_{i\in A_2} a_i \in W, \quad A_1, A_2 \supset A_0.$$

When X is complete, a family a is unconditionally summable in X if and only if the finite sums satisfy the latter Cauchy net condition. When X is complete and a_i, $i \in I$, is unconditionally summable in X, then for every subset $J \in I$, the corresponding subfamily a_j, $j \in J$, is also unconditionally summable in X.

When the sum of a family of non-negative numbers, in the extended sense defined before, is finite, then it coincides with the sum in the topological group $X = \mathbb{R}$.

If a family a in X is unconditionally summable, then for every W, neighbourhood of 0 in X, there is a finite subset A_0 of I such that $a_i \in W$ for every i not in A_0. If X is first-countable, it follows that the set of $i \in I$ such that $a_i \neq 0$ is countable. This need not be true in a general abelian topological group.

Unconditionally Convergent Series

Suppose that $I = \mathbb{N}$. If a family a_n, $n \in \mathbb{N}$, is unconditionally summable in an abelian Hausdorff topological group X, then the series in the usual sense converges and has the same sum,

$$\sum_{n=0}^{\infty} a_n = \sum_{n \in \mathbf{N}} a_n.$$

By nature, the definition of unconditional summability is insensitive to the order of the summation. When $\sum a_n$ is unconditionally summable, then the series remains convergent after any permutation σ of the set N of indices, with the same sum,

$$\sum_{n=0}^{\infty} a_{\sigma(n)} = \sum_{n=0}^{\infty} a_n.$$

Conversely, if every permutation of a series $\sum a_n$ converges, then the series is unconditionally convergent. When X is complete, then unconditional convergence is also equivalent to the fact that all subseries are convergent; if X is a Banach space, this is equivalent to say that for every sequence of signs $\varepsilon_n = 1$ or -1, the series

$$\sum_{n=0}^{\infty} \varepsilon_n a_n$$

converges in X. If X is a Banach space, then one may define the notion of absolute convergence. A series $\sum a_n$ of vectors in X converges absolutely if

$$\sum_{n \in \mathbf{N}} \| a_n \| < +\infty.$$

If a series of vectors in a Banach space converges absolutely then it converges unconditionally, but the converse only holds in finite-dimensional Banach spaces (theorem of Dvoretzky & Rogers (1950)).

Well-Ordered Sums

Conditionally convergent series can be considered if I is a well-ordered set, for example an ordinal number α_0. One may define by transfinite recursion:

$$\sum_{\beta<\alpha+1} a_\beta = a_\alpha + \sum_{\beta<\alpha} a_\beta$$

and for a limit ordinal α,

$$\sum_{\beta<\alpha} a_\beta = \lim_{\gamma\to\alpha} \sum_{\beta<\gamma} a_\beta$$

if this limit exists. If all limits exist up to α_0, then the series converges.

Examples

1. Given a function $f: X\to Y$, with Y an abelian topological group, define for every $a \in X$

$$f_a(x) = \begin{cases} 0 & x \neq a, \\ f(a) & x = a, \end{cases}$$

a function whose support is a singleton $\{a\}$. Then

$$f = \sum_{a\in X} f_a$$

in the topology of pointwise convergence (that is, the sum is taken in the infinite product group Y^X).

In the definition of partitions of unity, one constructs sums of functions over arbitrary index set I,

$$\sum_{i\in I} \varphi_i(x) = 1.$$

While, formally, this requires a notion of sums of uncountable series, by construction there are, for every given x, only finitely many nonzero terms in the sum, so issues regarding convergence of such sums do not arise. Actually, one usually assumes more: the family of functions is *locally finite, i.e.*, for every x there is a neighbourhood of x in which all but a finite number of functions vanish. Any regularity property of the φ_i, such as continuity, differentiability, that is preserved under finite sums will be preserved for the sum of any subcollection of this family of functions.

On the first uncountable ordinal ω_1 viewed as a topological space in the order topology, the constant function f: $[0,\omega_1) \to [0,\omega_1]$ given by $f(\alpha) = 1$ satisfies

$$\sum_{\alpha\in[0,\omega_1)} f(\alpha)=\omega_1$$

(in other words, ω_1 copies of 1 is ω_1) only if one takes a limit over all *countable* partial sums, rather than finite partial sums. This space is not separable.

Sets

In mathematics, a set is a collection of distinct objects, considered as an object in its own right. For example, the numbers 2, 4, and 6 are distinct objects when considered separately, but when they are considered collectively they form a single set of size three, written {2,4,6}. Sets are one of the most fundamental concepts in mathematics. Developed at the end of the 19th century, set theory is now a ubiquitous part of mathematics, and can be used as a foundation from which nearly all of mathematics can be derived. In mathematics education, elementary topics such as Venn diagrams are taught at a young age, while more advanced concepts are taught as part of a university degree. The German word *Menge*, rendered as "set" in English, was coined by Bernard Bolzano in his work The Paradoxes of the Infinite.

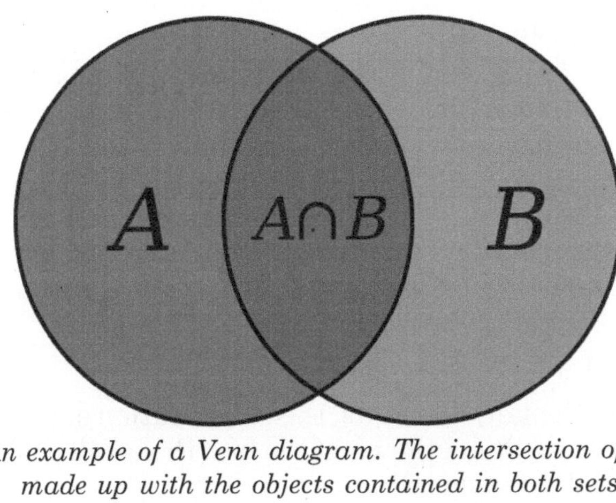

Figure: *An example of a Venn diagram. The intersection of two sets is made up with the objects contained in both sets*

A set is a well defined collection of distinct objects. The objects that make up a set (also known as the elements or members of a set) can be anything: numbers, people, letters of the alphabet, other sets, and so on. Georg Cantor, the founder of set theory, gave the following definition of a set at the beginning of his Beiträge zur Begründung der transfiniten Mengenlehre:

A set is a gathering together into a whole of definite, distinct objects of our perception [Anschauung] or of our thought—which are called elements of the set. Sets are conventionally denoted with capital letters. Sets A and B are equal if and only if they have precisely the same elements.

Cantor's definition turned out to be inadequate for formal mathematics; instead, the notion of a "set" is taken as an undefined primitive in axiomatic set theory, and its properties are defined by the Zermelo–Fraenkel axioms. The most basic properties are that a set has elements, and that two sets are equal (one and the same) if and only if every element of one set is an element of the other.

Describing Sets

There are two ways of describing, or specifying the members of, a set. One way is by intensional definition, using a rule or semantic description:

A is the set whose members are the first four positive integers.

B is the set of colours of the French flag.

The second way is by extension – that is, listing each member of the set. An extensional definition is denoted by enclosing the list of members in curly brackets:

$$C = \{4, 2, 1, 3\}$$

$$D = \{\text{blue, white, red}\}.$$

There are two important points to note about sets. First, a set can have two or more members which are identical, for example, {11, 6, 6}. However, we say that two sets which differ only in that one has duplicate members are in fact exactly identical. Hence, the set {11, 6, 6} is exactly identical to the set {11, 6}. The second important point is that the order in which the elements of a set are listed is irrelevant (unlike for a sequence or tuple). We can illustrate these two important points with an example:

$$\{6, 11\} = \{11, 6\} = \{11, 6, 6, 11\} .$$

For sets with many elements, the enumeration of members can be abbreviated. For instance, the set of the first thousand positive integers may be specified extensionally as:

$$\{1, 2, 3, ..., 1000\},$$

where the ellipsis ("...") indicates that the list continues in the obvious way. Ellipses may also be used where sets have infinitely many members. Thus the set of positive even numbers can be written as {2, 4, 6, 8, ... }.

The notation with braces may also be used in an intensional specification of a set. In this usage, the braces have the meaning "the set of all ...". So, E = {playing card suits} is the set whose four members are and c&. A more general form of this is set-builder notation, through which, for instance, the set F of the twenty smallest integers that are four less than perfect squares can be denoted:

$$F = \{n^2 - 4 : n \text{ is an integer; and } 0 \leq n \leq 19\}.$$

In this notation, the colon (":") means "such that", and the description can be interpreted as "F is the set of all numbers of the form $n^2 - 4$, such that n is a whole number in the range from 0 to 19 inclusive." Sometimes the vertical bar ("|") is used instead of the colon.

One often has the choice of specifying a set intensionally or extensionally. In the examples above, for instance, $A = C$ and $B = D$.

Membership

If a is a member of B, this is denoted $a \in B$, while if c is not a member of B then $c \notin B$. For example, with respect to the sets $A = \{1,2,3,4\}$, $B = \{\text{blue, white, red}\}$, and $F = \{n^2 - 4 : n \text{ is an integer; and } 0 \leq n \leq 19\}$ defined above,

$4 \in A$ and $12 \in F$; but

$9 \notin F$ and green $\notin B$.

Subsets

If every member of set A is also a member of set B, then A is said to be a *subset* of B, written $A \subseteq B$ (also pronounced *A is contained in B*). Equivalently, we can write $B \supseteq A$, read as *B is a superset of A*, *B includes A*, or *B contains A*. The relationship between sets established by †" is called *inclusion* or *containment*. If A is a subset of, but not equal to, B, then A is called a *proper subset* of B, written $A \subsetneq B$ (*A is a proper subset of B*) or $B \supsetneq A$ (*B is a proper superset of A*).

Note that the expressions $A \subset B$ and $B \supset A$ are used differently by different authors; some authors use them to mean the same as $A \subseteq B$ (respectively $B \supseteq A$), whereas other use them to mean the same as $A \subsetneq B$ (respectively $B \supsetneq A$).

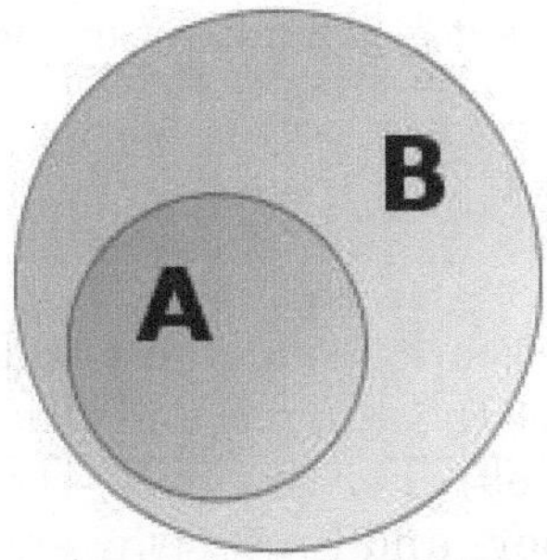

Figure: A *is a subset of* B

Example:

- The set of all men is a proper subset of the set of all people.
- $\{1, 3\} \subseteq \{1, 2, 3, 4\}$.
- $\{1, 2, 3, 4\} \subseteq \{1, 2, 3, 4\}$.

The empty set is a subset of every set and every set is a subset of itself:

- $\varnothing \subseteq A$.
- $A \subseteq A$.

An obvious but useful identity, which can often be used to show that two seemingly different sets are equal:

- $A = B$ if and only if $A \subseteq B$ and $B \subseteq A$.

A partition of a set S is a set of nonempty subsets of S such that every element x in S is in exactly one of these subsets.

Power Sets

The power set of a set S is the set of all subsets of S. Note that the power set contains S itself and the empty set because these are both subsets of S. For example, the power set of the set $\{1, 2, 3\}$ is $\{\{1, 2, 3\}, \{1, 2\}, \{1, 3\}, \{2, 3\}, \{1\}, \{2\}, \{3\}, \infty\}$. The power set of a set S is usually written as $P(S)$.

The power set of a finite set with n elements has 2^n elements. This relationship is one of the reasons for the terminology *power set*. For example, the set $\{1, 2, 3\}$ contains three elements, and the power set shown above contains $2^3 = 8$ elements.

The power set of an infinite (either countable or uncountable) set is always uncountable. Moreover, the power set of a set is always strictly "bigger" than the original set in the sense that there is no way to pair every element of S with exactly one element of $P(S)$. (There is never an onto map or surjection from S onto $P(S)$.)

Every partition of a set S is a subset of the powerset of S.

Cardinality

The cardinality $|S|$ of a set S is "the number of members of S." For example, if $B = \{blue, white, red\}$, $|B| = 3$.

There is a unique set with no members and zero cardinality, which is called the empty set (or the *null set*) and is denoted by the symbol $\varnothing$ (other notations are used). For example, the set of all three-sided squares has zero members and thus is the empty set. Though it may seem trivial, the empty set, like the number zero, is important in mathematics; indeed, the existence of this set is one of the fundamental concepts of axiomatic set theory.

Some sets have infinite cardinality. The set N of natural numbers, for instance, is infinite. Some infinite cardinalities are greater than others. For instance, the set of real numbers has greater cardinality than the set of natural numbers. However, it can be shown that the cardinality of (which is to say, the number of points on) a straight line is the same as the cardinality of any segment of that line, of the entire plane, and indeed of any finite-dimensional Euclidean space.

Special Sets

There are some sets that hold great mathematical importance and are referred to with such regularity that they have acquired special names and notational conventions to identify them. One of these is the empty set, denoted {}or $\varnothing$. Another is the unit set {x}, which contains exactly one element, namely x. Many of these sets are represented using blackboard bold or bold typeface. Special sets of numbers include:

- P or $\mathbb{P}$, denoting the set of all primes: P = {2, 3, 5, 7, 11, 13, 17, ...}.
- N or $\mathbb{N}$, denoting the set of all natural numbers: N = {1, 2, 3, . . .} (sometimes defined containing 0).
- Z or $\mathbb{Z}$, denoting the set of all integers (whether positive, negative or zero): Z = {..., –2, –1, 0, 1, 2, ...}.
- Q or $\mathbb{Q}$, denoting the set of all rational numbers (that is, the set of all proper and improper fractions): Q = $\{a/b : a, b \in$ Z, $b \neq 0\}$. For example, $1/4 \in$ Q and $11/6 \in$ Q. All integers are in this set since every integer *a* can be expressed as the fraction *a*/1 (Z $\subsetneq$ Q).
- R or $\mathbb{R}$, denoting the set of all real numbers. This set includes all rational numbers, together with all irrational numbers (that is, numbers that cannot be rewritten as fractions, such as $\sqrt{2}$, as well as transcendental numbers such as π, e and numbers that cannot be defined).
- C or $\mathbb{C}$, denoting the set of all complex numbers: C = $\{a + bi : a, b \in$ R}. For example, $1 + 2i \in$ C.
- H or $\mathbb{H}$, denoting the set of all quaternions: H = $\{a + bi + cj + dk : a, b, c, d \in$ R}. For example, $1 + i + 2j - k \in$ H.

Positive and negative sets are denoted by a superscript - or +, for example: $\mathbb{Q}^+$ represents the set of positive rational numbers.

Each of the above sets of numbers has an infinite number of elements, and each can be considered to be a proper subset of the sets listed below it. The primes are used less frequently than the others outside of number theory and related fields.

Basic Operations

There are several fundamental operations for constructing new sets from given sets.

Unions:

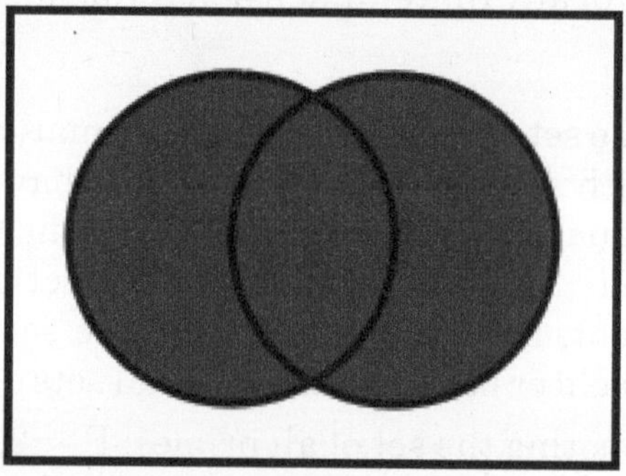

The union of A and B, denoted $A \cup B$

Two sets can be "added" together. The *union* of A and B, denoted by $A \cup B$, is the set of all things that are members of either A or B.

Examples:

- $\{1, 2\} \cup \{1, 2\} = \{1, 2\}$.
- $\{1, 2\} \cup \{2, 3\} = \{1, 2, 3\}$.
- $\{1, 2, 3\} \cup \{3, 4, 5\} = \{1, 2, 3, 4, 5\}$

Some basic properties of unions:

- $A \cup B = B \cup A$.
- $A \cup (B \cup C) = (A \cup B) \cup C$.
- $A \subseteq (A \cup B)$.
- $A \cup A = A$.
- $A \cup \varnothing = A$.
- $A \subseteq B$ if and only if $A \cup B = B$.

Intersections

A new set can also be constructed by determining which members two sets have "in common". The *intersection* of A and B, denoted by $A \cap B$, is the set of all things that are members of both A and B. If $A \cap B = \varnothing$, then A and B are said to be *disjoint*.

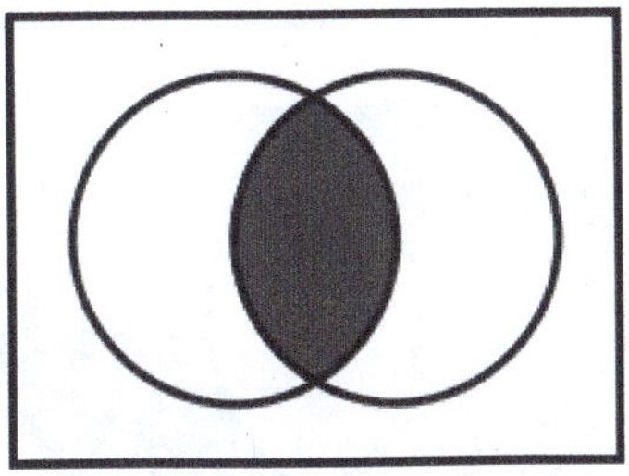

The intersection of A and B, denoted $A \cap B$.

Examples:

- $\{1, 2\} \cap \{1, 2\} = \{1, 2\}$.
- $\{1, 2\} \cap \{2, 3\} = \{2\}$.

Some basic properties of intersections:

- $A \cap B = B \cap A$.
- $A \cap (B \cap C) = (A \cap B) \cap C$.
- $A \cap B \subseteq A$.
- $A \cap A = A$.
- $A \cap \varnothing = \varnothing$.
- $A \subseteq B$ if and only if $A \cap B = A$.

Complements

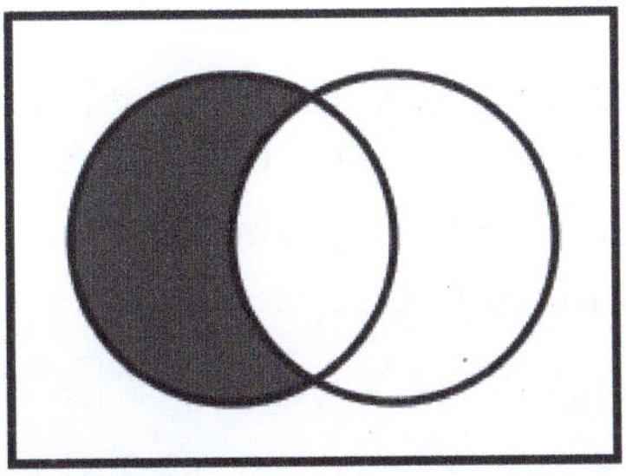

The relative complement of B in A

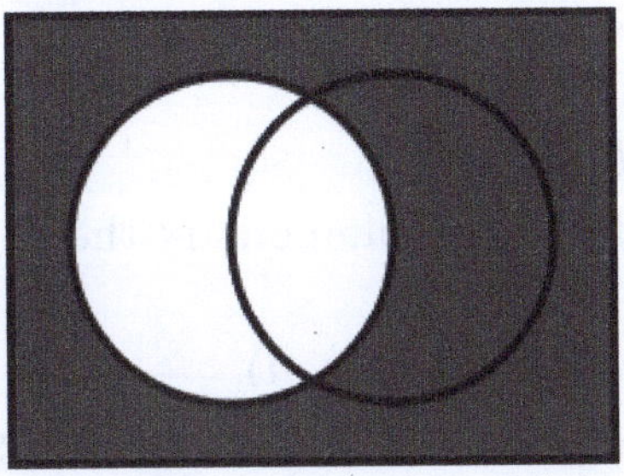

***Figure:** The complement of* A *in* U

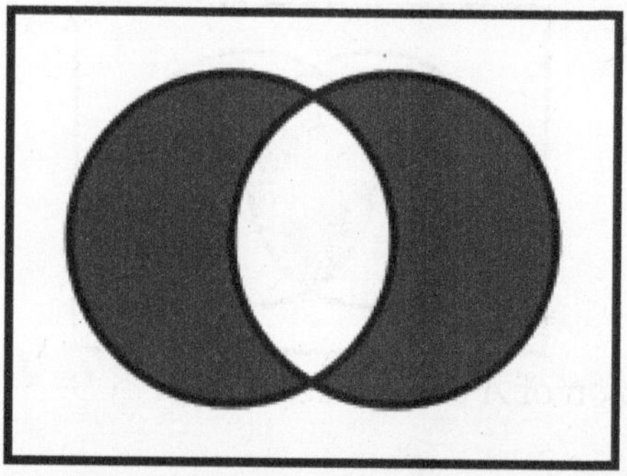

***Figure:**The symmetric difference of* A *and* B

Two sets can also be "subtracted". The *relative complement* of B in A (also called the *set-theoretic difference* of A and B), denoted by $A \setminus B$ (or $A - B$), is the set of all elements that are members of A but not members of B. Note that it is valid to "subtract" members of a set that are not in the set, such as removing the element *green* from the set {1, 2, 3}; doing so has no effect.

In certain settings all sets under discussion are considered to be subsets of a given universal set U. In such cases, $U \setminus A$ is called the *absolute complement* or simply *complement* of A, and is denoted by $A2$.

Examples

- $\{1, 2\} \setminus \{1, 2\} = \varnothing$.
- $\{1, 2, 3, 4\} \setminus \{1, 3\} = \{2, 4\}$.
- If U is the set of integers, E is the set of even integers, and O is the set of odd integers, then $U \setminus E = E' = O$.

Some basic properties of complements:

- $A \setminus B \neq B \setminus A$ for $A \neq B$.
- $A \varnothing A' = U$.
- $A \cap A2 = \varnothing$.
- $(A2)2 = A$.
- $A \setminus A = \varnothing$.
- $U2 = \varnothing$ and $\varnothing' = U$.
- $A \setminus B = A \cap B'$.

An extension of the complement is the symmetric difference, defined for sets A, B as

$$A \Delta B = (A \setminus B) \cup (B \setminus A).$$

For example, the symmetric difference of {7,8,9,10} and {9,10,11,12} is the set {7,8,11,12}.

Cartesian Product

A new set can be constructed by associating every element of one set with every element of another set. The *Cartesian product* of two sets A and B, denoted by $A \times B$ is the set of all ordered pairs (a, b) such that a is a member of A and b is a member of B.

Examples:

- $\{1, 2\} \times \{\text{red, white}\} = \{(1, \text{red}), (1, \text{white}), (2, \text{red}), (2, \text{white})\}$.
- $\{1, 2\} \times \{\text{red, white, green}\} = \{(1, \text{red}), (1, \text{white}), (1, \text{green}), (2, \text{red}), (2, \text{white}), (2, \text{green})\}$.
- $\{1, 2\} \times \{1, 2\} = \{(1, 1), (1, 2), (2, 1), (2, 2)\}$.

Some basic properties of cartesian products:

- $A \times \varnothing = \varnothing$.
- $A \times (B \cup C) = (A \times B) \cup (A \times C)$.
- $(A \cup B) \times C = (A \times C) \cup (B \times C)$.

Let A and B be finite sets. Then

- $|A \times B| = |B \times A| = |A| \times |B|$.

Applications

Set theory is seen as the foundation from which virtually all of mathematics can be derived. For example, structures in abstract algebra, such as groups, fields and rings, are sets closed under one or more operations.

One of the main applications of naive set theory is constructing relations. A relation from a domain A to a codomain B is a subset of the Cartesian product $A \times B$. Given this concept, we are quick to see that the set F of all ordered pairs (x, x^2), where x is real, is quite familiar. It has a domain set R and a codomain set that is also R, because the set of all squares is subset of the set of all reals. If placed in functional notation, this relation becomes $f(x) = x^2$. The reason these two are equivalent is for any given value, y that the function is defined for, its corresponding ordered pair, (y, y^2) is a member of the set F.

Axiomatic Set Theory

Although initially naive set theory, which defines a set merely as *any* well-defined collection, was well accepted, it soon ran into several obstacles. It was found that this definition spawned several paradoxes, most notably:

- Russell's paradox—It shows that the "set of all sets that *do not contain themselves*," i.e. the "set" $\{x : x \text{ is a set and } x - x\}$ does not exist.
- Cantor's paradox—It shows that "the set of all sets" cannot exist.

The reason is that the phrase *well-defined* is not very well defined. It was important to free set theory of these paradoxes because nearly all of mathematics was being redefined in terms of set theory. In an attempt to avoid these paradoxes, set theory was axiomatized based on first-order logic, and thus axiomatic set theory was born.

For most purposes however, naive set theory is still useful.

Principle of Inclusion and Exclusion

This principle gives us the cardinality of the union of sets.

$$\begin{aligned}|A_1 \cup A_2 \cup A_3 \cup \ldots \cup A_n| = \; & (|A_1| + |A_2| + |A_3| + \ldots |A_n|) - \\ & (|A_1 \cap A_2| + |A_1 \cap A_3| + \ldots |A_{n-1} \cap A_n|) + \\ & \ldots + \\ & (-1)^{n-1} (|A_1 \cap A_2 \cap A_3 \cap \ldots \cap A_n|)\end{aligned}$$

De Morgan's Law

De Morgan stated two laws about Sets.

If A and B are any two Sets then,

- (A ∪ B)' = A' ∩ B'

The complement of A union B equals the complement of A intersected with the complement of B.

- (A ∩ B)' = A' ∪ B'

The complement of A intersected with B is equal to the complement of A union to the complement of B.

Chapter 5

Relations and Functions

In relational database theory, a relation, as originally defined by E. F. Codd, is a set of tuples (d_1, d_2, ..., d_n), where each element d_j is a member of D_j, a data domain. Codd's original definition notwithstanding, and contrary to the usual definition in mathematics, there is no ordering to the elements of the tuples of a relation. Instead, each element is termed an attribute value. An attribute is a name paired with a domain (nowadays more commonly referred to as type or data type). An attribute value is an attribute name paired with an element of that attribute's domain, and a tuple is a *set* of attribute values in which no two distinct elements have the same name. Thus, in some accounts, a tuple is described as a function, mapping names to values.

A set of attributes in which no two distinct elements have the same name is called a heading. A set of tuples having the same heading is called a body. A relation is thus a heading paired with a body, the heading of the relation being also the heading of each tuple in its body. The number of attributes constituting a heading is called the degree, which term also applies to tuples and relations. The term *n*-tuple refers to a tuple of degree *n* ($n>=0$).

E. F. Codd used the term relation in its mathematical sense of a finitary relation, a set of tuples on some set of *n* sets S_1, S_2, ... ,S_n. Thus, an *n*-ary relation is interpreted, under the Closed World Assumption, as the extension of some *n*-adic predicate: all and only those *n*-tuples whose values, substituted for corresponding free variables in the predicate, yield propositions that hold true, appear in the relation.

The term relation schema refers to a heading paired with a set of constraints defined in terms of that heading. A relation can thus be seen as an instantiation of a relation schema if it has the heading of that schema and it satisfies the applicable constraints.

Sometimes a relation schema is taken to include a name. A relational database definition (database schema, sometimes referred to as a relation*al* schema) can thus be thought of as a collection of named relation schemas.

In implementations, the domain of each attribute is effectively a data type and a named relation schema is effectively a relation variable or relvar for short.

In SQL, a database language for relational databases, relations are represented by tables, where each row of a table represents a single tuple, and where the values of each attribute form a column.

Examples

Below is an example of a relation having three named attributes: 'ID' from the domain of integers, and 'Name' and 'Address' from the domain of strings:

ID (Integer)	*Name (String)*	*Address (String)*
102	Yonezawa Akinori	Naha, Okinawa
202	Murata Makoto	Sendai, Miyagi
104	Sakamura Ken	Kumamoto, Kumamoto
152	Matsumoto Yukihiro	Okinawa, Okinawa

A predicate for this relation, using the attribute names to denote free variables, might be "Employee number *ID* is known as *Name* and lives at *Address*". Examination of the relation tells us that there are just four tuples for which the predicate holds true. So, for example, employee 102 is known only by that name, Yonezawa Akinori, and does not live anywhere else but in Naha, Okinawa. Also, apart from the four employees shown, there is no other employee who has both a name and an address.

Under the definition of body, the tuples of a body do not appear in any particular order - one cannot say "The tuple of 'Murata Makoto' is above the tuple of 'Matsumoto Yukihiro'", nor can one say "The tuple of 'Yonezawa Akinori' is the first tuple." A similar comment applies to the rows of an SQL table.

Under the definition of heading, the attributes of an element do not appear in any particular order either, nor, therefore do the elements of a tuple. A similar comment does *not* apply here to SQL, which does define an ordering to the columns of a table.

Relation Variables

A relational database consists of named relation variables (relvars) for the purposes of updating the database in response to changes in the real world. An update to a single relvar causes the body of the relation assigned to that variable to be replaced by a different set of tuples. Such variables are classified into two classes: base relation variables and derived relation variables, the latter also known as virtual relvars but usually referred to by the short term view.

A base relation variable is a relation variable which is not derived from any other relation variables. In SQL the term base table equates approximately to base relation variable.

A view can be defined by an expression using the operators of the relational algebra or the relational calculus. Such an expression operates on one or more relations and when evaluated yields another relation. The result is sometimes referred to as a "derived" relation when the operands are relations assigned to database variables. A view is defined by giving a name to such an expression, such that the name can subsequently be used as a variable name.

Correlation and Dependence

In statistics, dependence is any statistical relationship between two random variables or two sets of data. Correlation refers to any of a broad class of statistical relationships involving dependence.

Familiar examples of dependent phenomena include the correlation between the physical statures of parents and their offspring, and the correlation between the demand for a product and its price. Correlations are useful because they can indicate a predictive relationship that can be exploited in practice. For example, an electrical utility may produce less power on a mild day based on the correlation between electricity demand and weather. In this example there is a causal relationship, because extreme weather causes people to use more electricity for heating or cooling; however, statistical dependence is not sufficient to demonstrate the presence of such a causal relationship (i.e., correlation does not imply causation).

Formally, *dependence* refers to any situation in which random variables do not satisfy a mathematical condition of probabilistic independence. In loose usage, *correlation* can refer to any departure of two or more random variables from independence, but technically it refers to any of several more specialised types of relationship between mean values. There are several correlation coefficients, often denoted

ρ or r, measuring the degree of correlation. The most common of these is the Pearson correlation coefficient, which is sensitive only to a linear relationship between two variables (which may exist even if one is a nonlinear function of the other). Other correlation coefficients have been developed to be more robust than the Pearson correlation – that is, more sensitive to nonlinear relationships. Mutual information can also be applied to measure dependence between two variables.

Limits and Continuity

In mathematics, a limit is the value that a function or sequence "approaches" as the input or index approaches some value. Limits are essential to calculus (and mathematical analysis in general) and are used to define continuity, derivatives, and integrals.

The concept of a limit of a sequence is further generalized to the concept of a limit of a topological net, and is closely related to limit and direct limit in category theory.

In formulas, a limit is usually denoted "lim" as in $\lim_{n \to c}(a_n) = L$, and the fact of approaching a limit is represented by the right arrow ($\to$) as in $a_n \to L$.

Limit of a Function

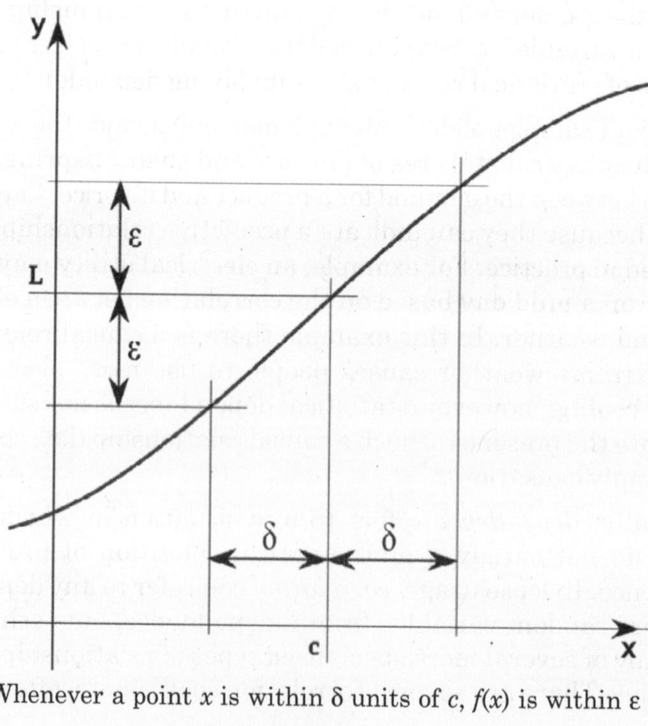

Figure: Whenever a point x is within δ units of c, $f(x)$ is within ε units of L.

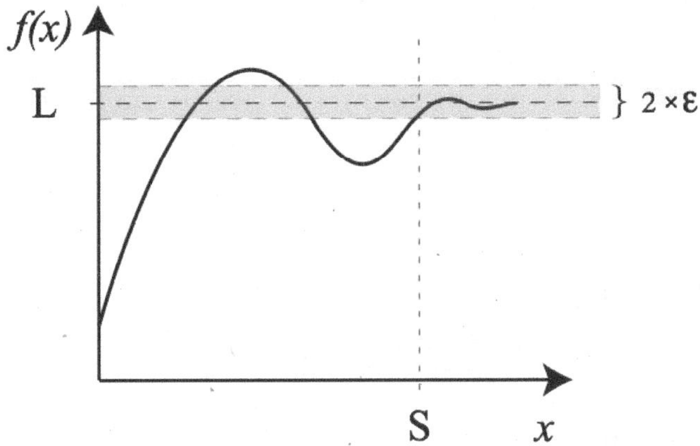

Figure: *For all $x > S$, $f(x)$ is within ε of L.*

Suppose f is a real-valued function and c is a real number. The expression

$$\lim_{x \to c} f(x) = L$$

means that $f(x)$ can be made to be as close to L as desired by making x sufficiently close to c. In that case, the above equation can be read as "the limit of f of x, as x approaches c, is L".

Augustin-Louis Cauchy in 1821, followed by Karl Weierstrass, formalized the definition of the limit of a function as the above definition, which became known as the (ε, δ)-definition of limit in the 19th century. The definition uses ε (the lowercase Greek letter *epsilon*) to represent any small positive number, so that –$f(x)$ becomes arbitrarily close to L" means that $f(x)$ eventually lies in the interval $(L - \varepsilon, L + \varepsilon)$, which can also be written using the absolute value sign as $|f(x) - L| < \varepsilon$. The phrase "as x approaches c– then indicates that we refer to values of x whose distance from c is less than some positive number δ (the lower case Greek letter *delta*)—that is, values of x within either $(c - \delta, c)$ or $(c, c + \delta)$, which can be expressed with $0 < |x - c| < \delta$. The first inequality means that the distance between x and c is greater than 0 and that $x \neq c$, while the second indicates that x is within distance δ of c.

Note that the above definition of a limit is true even if $f(c) \neq L$. Indeed, the function f need not even be defined at c.

For example, if

$$f(x) = \frac{x^2 - 1}{x - 1}$$

then $f(1)$ is not defined, yet as x moves arbitrarily close to 1, $f(x)$ correspondingly approaches 2:

$f(0.9)$	$f(0.99)$	$f(0.999)$		$f(1.0)$		$f(1.001)$	$f(1.01)$	$f(1.1)$
1.900	1.990	1.999	$\Rightarrow$	undefined	$\Leftarrow$	2.001	2.010	2.100

Thus, $f(x)$ can be made arbitrarily close to the limit of 2 just by making x sufficiently close to 1.

In other words, $\lim_{x\to 1}\frac{x^2-1}{x-1}=2$

This can also be calculated algebraically, as

$$\frac{x^2-1}{x-1}=\frac{(x+1)(x-1)}{x-1}=x+1$$

for all real numbers $x \neq 1$.

Now since $x + 1$ is continuous in x at 1, we can now plug in 1 for x, thus $\lim_{x\to 1}\frac{x^2-1}{x-1}=1+1=2$.

In addition to limits at finite values, functions can also have limits at infinity. For example, consider

$$f(x)=\frac{2x-1}{x}$$

- $f(100) = 1.9900$
- $f(1000) = 1.9990$
- $f(10000) = 1.99990$

As x becomes extremely large, the value of $f(x)$ approaches 2, and the value of $f(x)$ can be made as close to 2 as one could wish just by picking x sufficiently large. In this case, the limit of $f(x)$ as x approaches infinity is 2. In mathematical notation,

$$\lim_{x\to\infty}\frac{2x-1}{x}=2.$$

Limit of a Sequence

Consider the following sequence: 1.79, 1.799, 1.7999,... It can be observed that the numbers are "approaching" 1.8, the limit of the sequence.

Formally, suppose a_1, a_2, ... is a sequence of real numbers. It can be stated that the real number L is the *limit* of this sequence, namely:

$$\lim_{n\to\infty} a_n = L$$

to mean.

For every real number $\varepsilon > 0$, there exists a natural number N such that for all $n > N$, we have $|a_n - L| < \varepsilon$.

Intuitively, this means that eventually all elements of the sequence get arbitrarily close to the limit, since the absolute value $|a_n - L|$ is the distance between a_n and L. Not every sequence has a limit; if it does, it is called convergent, and if it does not, it is *divergent*. One can show that a convergent sequence has only one limit.

The limit of a sequence and the limit of a function are closely related. On one hand, the limit as n goes to infinity of a sequence $a(n)$ is simply the limit at infinity of a function defined on the natural numbers n. On the other hand, a limit L of a function $f(x)$ as x goes to infinity, if it exists, is the same as the limit of any arbitrary sequence a_n that approaches L, and where a_n is never equal to L. Note that one such sequence would be $L + 1/n$.

Limit as "Standard Part"

In non-standard analysis (which involves a hyperreal enlargement of the number system), the limit of a sequence (a_n) can be expressed as the standard part of the value a_H of the natural extension of the sequence at an infinite hypernatural index n=H. Thus,

$$\lim_{n\to\infty} a_n = \mathrm{st}(a_H).$$

Here the standard part function "st" rounds off each finite hyperreal number to the nearest real number (the difference between them is infinitesimal). This formalizes the natural intuition that for "very large" values of the index, the terms in the sequence are "very close" to the limit value of the sequence. Conversely, the standard part of a hyperreal $a = [a_n]$ represented in the ultrapower construction by a Cauchy sequence (a_n), is simply the limit of that sequence:

$$\mathrm{st}(a) = \lim_{n\to\infty} a_n.$$

In this sense, taking the limit and taking the standard part are equivalent procedures.

Differential Calculus

In mathematics, differential calculus is a subfield of calculus concerned with the study of the rates at which quantities change. It is one of the two traditional divisions of calculus, the other being integral calculus.

The primary objects of study in differential calculus are the derivative of a function, related notions such as the differential, and their applications. The derivative of a function at a chosen input value describes the rate of change of the function near that input value. The process of finding a derivative is called differentiation. Geometrically, the derivative at a point is the slope of the tangent line to the graph of the function at that point, provided that the derivative exists and is defined at that point. For a real-valued function of a single real variable, the derivative of a function at a point generally determines the best linear approximation to the function at that point.

Differential calculus and integral calculus are connected by the fundamental theorem of calculus, which states that differentiation is the reverse process to integration.

Differentiation has applications to nearly all quantitative disciplines. For example, in physics, the derivative of the displacement of a moving body with respect to time is the velocity of the body, and the derivative of velocity with respect to time is acceleration. Newton's second law of motion states that the derivative of the momentum of a body equals the force applied to the body. The reaction rate of a chemical reaction is a derivative. In operations research, derivatives determine the most efficient ways to transport materials and design factories.

Derivatives are frequently used to find the maxima and minima of a function. Equations involving derivatives are called differential equations and are fundamental in describing natural phenomena. Derivatives and their generalizations appear in many fields of mathematics, such as complex analysis, functional analysis, differential geometry, measure theory and abstract algebra.

The Derivative

Suppose that x and y are real numbers and that y is a function of x, that is, for every value of x, there is a corresponding value of y. This relationship can be written as $y = f(x)$. If $f(x)$ is the equation for a straight line, then there are two real numbers m and b such that $y = m\,x + b$. m is called the slope and can be determined from the formula:

$$m = \frac{\text{change in } y}{\text{change in } x} = \frac{\Delta y}{\Delta x},$$

where the symbol Δ (the uppercase form of the Greek letter Delta) is an abbreviation for "change in". It follows that $\Delta y = m\,\Delta x$.

A general function is not a line, so it does not have a slope. The derivative of f at the point x is the slope of the linear approximation to *f* at the point x. It is usually denoted $f'(x)$ or dy/dx. Together with the value of f at x, the derivative of f determines the best linear approximation, or linearization, of f near the point x. This latter property is usually taken as the definition of the derivative.

A closely related notion is the differential of a function.

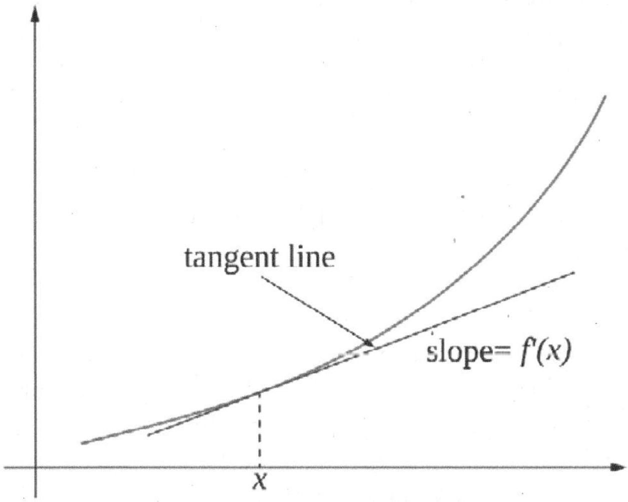

***Figure:** The tangent line at (x,f(x))*

When x and y are real variables, the derivative of f at x is the slope of the tangent line to the graph of f at x. Because the source and target of f are one-dimensional, the derivative of f is a real number. If x and y are vectors, then the best linear approximation to the graph of f depends on how f changes in several directions at once. Taking the best linear approximation in a single direction determines a partial derivative, which is usually denoted $\partial y / \partial x$. The linearization of f in all directions at once is called the total derivative.

History of Differentiation

The concept of a derivative in the sense of a tangent line is a very old one, familiar to Greek geometers such as Euclid (c. 300 BC), Archimedes (c. 287–212 BC) and Apollonius of Perga (c. 262–190 BC). Archimedes also introduced the use of infinitesimals, although these were primarily used to study areas and volumes rather than derivatives and tangents.

The use of infinitesimals to study rates of change can be found in Indian mathematics, perhaps as early as 500 AD, when the astronomer and mathematician Aryabhata (476–550) used infinitesimals to study the motion of the moon. The use of infinitesimals to compute rates of change was developed significantly by Bhâskara II (1114–1185); indeed, it has been argued[3] that many of the key notions of differential calculus can be found in his work, such as "Rolle's theorem". The Persian mathematician, Sharaf al-Dîn al-Tûsî (1135–1213), was the first to discover the derivative of cubic polynomials, an important result in differential calculus; his *Treatise on Equations* developed concepts related to differential calculus, such as the derivative function and the maxima and minima of curves, in order to solve cubic equations which may not have positive solutions.

The modern development of calculus is usually credited to Isaac Newton (1643–1727) and Gottfried Leibniz (1646–1716), who provided independent and unified approaches to differentiation and derivatives. The key insight, however, that earned them this credit, was the fundamental theorem of calculus relating differentiation and integration: this rendered obsolete most previous methods for computing areas and volumes, which had not been significantly extended since the time of Ibn al-Haytham (Alhazen). For their ideas on derivatives, both Newton and Leibniz built on significant earlier work by mathematicians such as Isaac Barrow (1630–1677), René Descartes (1596–1650), Christiaan Huygens (1629–1695), Blaise Pascal (1623–1662) and John Wallis (1616–1703). Isaac Barrow is generally given credit for the early development of the derivative. Nevertheless, Newton and Leibniz remain key figures in the history of differentiation, not least because Newton was the first to apply differentiation to theoretical physics, while Leibniz systematically developed much of the notation still used today.

Since the 17th century many mathematicians have contributed to the theory of differentiation. In the 19th century, calculus was put on a much more rigorous footing by mathematicians such as Augustin Louis Cauchy (1789–1857), Bernhard Riemann (1826–1866), and Karl Weierstrass (1815–1897). It was also during this period that the differentiation was generalized to Euclidean space and the complex plane.

Applications of Derivatives

Optimization: If f is a differentiable function on R (or an open interval) and x is a local maximum or a local minimum of f, then the derivative of f at x is zero; points where $f'(x) = 0$ are called critical points or stationary points (and the value of f at x is called a critical value). (The definition of a critical point is sometimes extended to include points where the derivative does not exist.) Conversely, a critical point x of f can be analysed by considering the second derivative of f at x:

- if it is positive, x is a local minimum;
- if it is negative, x is a local maximum;
- if it is zero, then x could be a local minimum, a local maximum, or neither. (For example, $f(x) = x^3$ has a critical point at $x = 0$, but it has neither a maximum nor a minimum there, whereas $f(x) = \pm x^4$ has a critical point at $x = 0$ and a minimum and a maximum, respectively, there.)

This is called the second derivative test. An alternative approach, called the first derivative test, involves considering the sign of the f' on each side of the critical point.

Taking derivatives and solving for critical points is therefore often a simple way to find local minima or maxima, which can be useful in optimization. By the extreme value theorem, a continuous function on a closed interval must attain its minimum and maximum values at least once. If the function is differentiable, the minima and maxima can only occur at critical points or endpoints.

This also has applications in graph sketching: once the local minima and maxima of a differentiable function have been found, a rough plot of the graph can be obtained from the observation that it will be either increasing or decreasing between critical points.

In higher dimensions, a critical point of a scalar valued function is a point at which the gradient is zero. The second derivative test can still be used to analyse critical points by considering the eigenvalues of the Hessian matrix of second partial derivatives of the function at the critical point. If all of the eigenvalues are positive, then the point is a local minimum; if all are negative, it is a local maximum. If there are some positive and some negative eigenvalues, then the critical point is a saddle point, and if none of these cases hold (i.e., some of the eigenvalues are zero) then the test is inconclusive.

Calculus of Variations

One example of an optimization problem is: Find the shortest curve between two points on a surface, assuming that the curve must also lie on the surface. If the surface is a plane, then the shortest curve is a line. But if the surface is, for example, egg-shaped, then the shortest path is not immediately clear. These paths are called geodesics, and one of the simplest problems in the calculus of variations is finding geodesics. Another example is: Find the smallest area surface filling in a closed curve in space. This surface is called a minimal surface and it, too, can be found using the calculus of variations.

Physics

Calculus is of vital importance in physics: many physical processes are described by equations involving derivatives, called differential equations. Physics is particularly concerned with the way quantities change and evolve over time, and the concept of the "time derivative" — the rate of change over time — is essential for the precise definition of several important concepts. In particular, the time derivatives of an object's position are significant in Newtonian physics:

- velocity is the derivative (with respect to time) of an object's displacement (distance from the original position)
- acceleration is the derivative (with respect to time) of an object's velocity, that is, the second derivative (with respect to time) of an object's position.

For example, if an object's position on a line is given by

$$x(t) = -16t^2 + 16t + 32,$$

then the object's velocity is

$$\dot{x}(t) = x'(t) = -32t + 16,$$

and the object's acceleration is

$$\ddot{x}(t) = x''(t) = -32,$$

which is constant.

Differential Equations

A differential equation is a relation between a collection of functions and their derivatives. An ordinary differential equation is a differential equation that relates functions of one variable to their derivatives with respect to that variable. A partial differential equation is a differential equation that relates functions of more than one variable to their partial derivatives. Differential equations arise naturally in the physical sciences, in mathematical modelling, and within mathematics itself. For example, Newton's second law, which describes the relationship between acceleration and force, can be stated as the ordinary differential equation

$$F(t) = m\frac{d^2x}{dt^2}.$$

The heat equation in one space variable, which describes how heat diffuses through a straight rod, is the partial differential equation

$$\frac{\partial u}{\partial t} = \alpha\frac{\partial^2 u}{\partial x^2}.$$

Here $u(x,t)$ is the temperature of the rod at position x and time t and α is a constant that depends on how fast heat diffuses through the rod.

Mean Value Theorem

The mean value theorem gives a relationship between values of the derivative and values of the original function. If $f(x)$ is a real-valued function and a and b are numbers with $a < b$, then the mean value theorem says that under mild hypotheses, the slope between the two points $(a,f(a))$ and $(b,f(b))$ is equal to the slope of the tangent line to f at some point c between a and b. In other words,

$$f'(c) = \frac{f(b) - f(a)}{b - a}.$$

In practice, what the mean value theorem does is control a function in terms of its derivative. For instance, suppose that f has derivative equal to zero at each point. This means that its tangent line is horizontal at every point, so the function should also be horizontal. The mean value theorem proves that this must be true: The slope between any two points on the graph of f must equal the slope of one of the tangent lines of f. All of those slopes are zero, so any line from one point on the graph to another point will also have slope zero. But that says that the function does not move up or down, so it must be a horizontal line. More complicated conditions on the derivative lead to less precise but still highly useful information about the original function.

Taylor Polynomials and Taylor Series

The derivative gives the best possible linear approximation, but this can be very different from the original function. One way of improving the approximation is to take a quadratic approximation. That is to say, the linearization of a real-valued function $f(x)$ at the point x_0 is a linear polynomial $a + b(x - x_0)$, and it may be possible to get a better approximation by considering a quadratic polynomial $a + b(x - x_0) + c(x - x_0)^2$. Still better might be a cubic polynomial $a + b(x - x_0) + c(x - x_0)^2 + d(x - x_0)^3$, and this idea can be extended to arbitrarily high degree polynomials. For each one of these polynomials, there should be a best possible choice of coefficients a, b, c, and d that makes the approximation as good as possible.

In the neighbourhood of x_0, for a the best possible choice is always $f(x_0)$, and for b the best possible choice is always $f'(x_0)$. For c, d, and higher-degree coefficients, these coefficients are determined by higher derivatives of f. c should always be $f''(x_0)/2$, and d should always be

$f'''(x_0)/3!$. Using these coefficients gives the Taylor polynomial of f. The Taylor polynomial of degree d is the polynomial of degree d which best approximates f, and its coefficients can be found by a generalization of the above formulas. Taylor's theorem gives a precise bound on how good the approximation is. If f is a polynomial of degree less than or equal to d, then the Taylor polynomial of degree d equals f.

The limit of the Taylor polynomials is an infinite series called the Taylor series. The Taylor series is frequently a very good approximation to the original function. Functions which are equal to their Taylor series are called analytic functions. It is impossible for functions with discontinuities or sharp corners to be analytic, but there are smooth functions which are not analytic.

Implicit Function Theorem

Some natural geometric shapes, such as circles, cannot be drawn as the graph of a function. For instance, if $f(x,y) = x^2 + y^2 - 1$, then the circle is the set of all pairs (x,y) such that $f(x,y) = 0$. This set is called the zero set of f. It is not the same as the graph of f, which is a cone. The implicit function theorem converts relations such as $f(x,y) = 0$ into functions. It states that if f is continuously differentiable, then around most points, the zero set of f looks like graphs of functions pasted together. The points where this is not true are determined by a condition on the derivative of f. The circle, for instance, can be pasted together from the graphs of the two functions $\pm\sqrt{1-x^2}$. In a neighbourhood of every point on the circle except (–1,0) and (1,0), one of these two functions has a graph that looks like the circle. (These two functions also happen to meet (–1,0) and (1,0), but this is not guaranteed by the implicit function theorem.)

The implicit function theorem is closely related to the inverse function theorem, which states when a function looks like graphs of invertible functions pasted together.

Integral Calculus

The integral is an important concept in mathematics. Integration, together with its inverse, differentiation, is one of the two main operations in calculus. Given a function f of a real variable x and an interval $[a, b]$ of the real line, the definite integral

$$\int_a^b f(x)dx$$

is defined informally as the signed area of the region in the xy-plane that is bounded by the graph of f, the x-axis and the vertical lines $x = a$

and $x = b$. The area above the x-axis adds to the total and that below the x-axis subtracts from the total.

The term *integral* may also refer to the related notion of the antiderivative, a function F whose derivative is the given function f. In this case, it is called an indefinite integral and is written:

$$F(x) = \int f(x)dx.$$

However, the integrals discussed in this article are those termed *definite integrals*.

The principles of integration were formulated independently by Isaac Newton and Gottfried Leibniz in the late 17th century. Through the fundamental theorem of calculus, which they independently developed, integration is connected with differentiation: if f is a continuous real-valued function defined on a closed interval $[a, b]$, then, once an antiderivative F of f is known, the definite integral of f over that interval is given by

$$\int_a^b f(x)dx = F(b) - F(a).$$

Integrals and derivatives became the basic tools of calculus, with numerous applications in science and engineering. The founders of calculus thought of the integral as an infinite sum of rectangles of infinitesimal width. A rigorous mathematical definition of the integral was given by Bernhard Riemann. It is based on a limiting procedure which approximates the area of a curvilinear region by breaking the region into thin vertical slabs. Beginning in the nineteenth century, more sophisticated notions of integrals began to appear, where the type of the function as well as the domain over which the integration is performed has been generalised. A line integral is defined for functions of two or three variables, and the interval of integration $[a, b]$ is replaced by a certain curve connecting two points on the plane or in the space. In a surface integral, the curve is replaced by a piece of a surface in the three-dimensional space.

Integrals of differential forms play a fundamental role in modern differential geometry. These generalizations of integrals first arose from the needs of physics, and they play an important role in the formulation of many physical laws, notably those of electrodynamics. There are many modern concepts of integration, among these, the most common is based on the abstract mathematical theory known as Lebesgue integration, developed by Henri Lebesgue.

Pre-Calculus Integration

The first documented systematic technique capable of determining integrals is the method of exhaustion of the ancient Greek astronomer Eudoxus (*ca.* 370 BC), which sought to find areas and volumes by breaking them up into an infinite number of divisions for which the area or volume was known. This method was further developed and employed by Archimedes in the 3rd century BC and used to calculate areas for parabolas and an approximation to the area of a circle. Similar methods were independently developed in China around the 3rd century AD by Liu Hui, who used it to find the area of the circle. This method was later used in the 5th century by Chinese father-and-son mathematicians Zu Chongzhi and Zu Geng to find the volume of a sphere (Shea 2007; Katz 2004, pp. 125–126).

The next significant advances in integral calculus did not begin to appear until the 16th century. At this time the work of Cavalieri with his method of indivisibles, and work by Fermat, began to lay the foundations of modern calculus, with Cavalieri computing the integrals of x^n up to degree $n = 9$ in Cavalieri's quadrature formula. Further steps were made in the early 17th century by Barrow and Torricelli, who provided the first hints of a connection between integration and differentiation. Barrow provided the first proof of the fundamental theorem of calculus. Wallis generalized Cavalieri's method, computing integrals of x to a general power, including negative powers and fractional powers.

Newton and Leibniz

The major advance in integration came in the 17th century with the independent discovery of the fundamental theorem of calculus by Newton and Leibniz. The theorem demonstrates a connection between integration and differentiation. This connection, combined with the comparative ease of differentiation, can be exploited to calculate integrals. In particular, the fundamental theorem of calculus allows one to solve a much broader class of problems. Equal in importance is the comprehensive mathematical framework that both Newton and Leibniz developed. Given the name infinitesimal calculus, it allowed for precise analysis of functions within continuous domains. This framework eventually became modern calculus, whose notation for integrals is drawn directly from the work of Leibniz.

Formalization

While Newton and Leibniz provided a systematic approach to integration, their work lacked a degree of rigour. Bishop Berkeley memorably attacked the vanishing increments used by Newton, calling them "ghosts of departed quantities". Calculus acquired a firmer footing with the development of limits.

Integration was first rigorously formalized, using limits, by Riemann. Although all bounded piecewise continuous functions are Riemann integrable on a bounded interval, subsequently more general functions were considered—particularly in the context of Fourier analysis—to which Riemann's definition does not apply, and Lebesgue formulated a different definition of integral, founded in measure theory (a subfield of real analysis).

Other definitions of integral, extending Riemann's and Lebesgue's approaches, were proposed. These approaches based on the real number system are the ones most common today, but alternative approaches exist, such as a definition of integral as the standard part of an infinite Riemann sum, based on the hyperreal number system.

Historical Notation

Isaac Newton used a small vertical bar above a variable to indicate integration, or placed the variable inside a box. The vertical bar was easily confused with $\dot{x}$ or x', which Newton used to indicate differentiation, and the box notation was difficult for printers to reproduce, so these notations were not widely adopted.

The modern notation for the indefinite integral was introduced by Gottfried Leibniz in 1675 (Burton 1988, p. 359; Leibniz 1899, p. 154). He adapted the integral symbol, $\int$, from the letter *n* (long s), standing for *summa* (written as *numma*; Latin for "sum" or "total"). The modern notation for the definite integral, with limits above and below the integral sign, was first used by Joseph Fourier in *Mémoires* of the French Academy around 1819–20, reprinted in his book of 1822 (Cajori 1929, pp. 249–250; Fourier 1822, §231).

Terminology and Notation

The simplest case, the integral over *x* of a real-valued function *f*(*x*), is written as

$$\int f(x)dx.$$

The integral sign +" represents integration. The *dx* indicates that we are integrating over *x*; *x* is called the variable of integration. Inside the $\int ...dx$ is the expression to be integrated, called the *integrand*. In correct mathematical typography, the *dx* is separated from the integrand by a space (as shown). Some authors use an upright *d* (that is, dx instead of *dx*). In this case the integrand is the function *f*(*x*). Because there is no domain specified, the integral is called an *indefinite integral.*

When integrating over a specified domain, we speak of a *definite integral.* Integrating over a domain *D* is written as $\int_D f(x)dx$, or $\int_a^b f(x)dx$ if the domain is an interval [*a*, *b*] of *x*.

The domain *D* or the interval [*a*, *b*] is called the *domain of integration.*

If a function has an integral, it is said to be *integrable.* In general, the integrand may be a function of more than one variable, and the domain of integration may be an area, volume, a higher-dimensional region, or even an abstract space that does not have a geometric structure in any usual sense (such as a sample space in probability theory).

In modern Arabic mathematical notation, a reflected integral symbol ∫ is used (W3C 2006).

The variable of integration *dx* has different interpretations depending on the theory being used. It can be seen as strictly a notation indicating that *x* is a dummy variable of integration; if the integral is seen as a Riemann sum, *dx* is a reflection of the weights or widths *d* of the intervals of *x*; in Lebesgue integration and its extensions, *dx* is a measure; in non-standard analysis, it is an infinitesimal; or it can be seen as an independent mathematical quantity, a differential form. More complicated cases may vary the notation slightly. In Leibniz's notation, *dx* is interpreted as an infinitesimal change in *x*. Although Leibniz's interpretation lacks rigour, his integration notation is the most common one in use today.

Introduction

Integrals appear in many practical situations. If a swimming pool is rectangular with a flat bottom, then from its length, width, and depth we can easily determine the volume of water it can contain (to fill it), the area of its surface (to cover it), and the length of its edge (to rope it). But if it is oval with a rounded bottom, all of these quantities call for integrals. Practical approximations may suffice for such trivial examples, but precision engineering (of any discipline) requires exact and rigorous values for these elements.

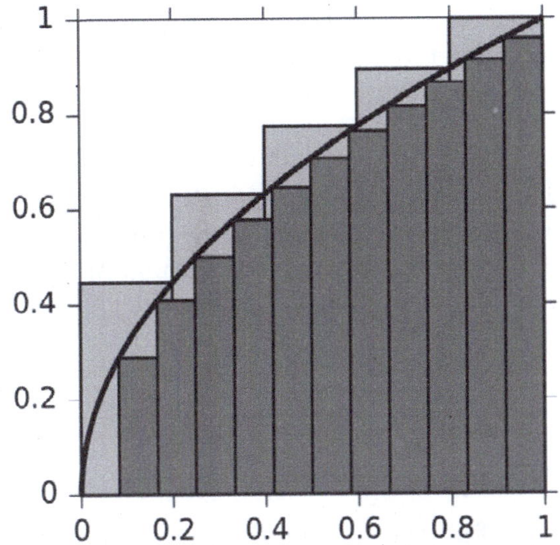

Figure: *Approximations to integral of √x from 0 to 1, with 5 right endpoint partitions and 12 left endpoint partitions*

To start off, consider the curve $y = f(x)$ between $x = 0$ and $x = 1$ with $f(x) = \sqrt{x}$. We ask:

What is the area under the function f, in the interval from 0 to 1?

and call this (yet unknown) area the integral of f. The notation for this integral will be

$$\int_0^1 \sqrt{x}\, dx.$$

As a first approximation, look at the unit square given by the sides $x = 0$ to $x = 1$ and $y = f(0) = 0$ and $y = f(1) = 1$. Its area is exactly 1. As it is, the true value of the integral must be somewhat less. Decreasing the width of the approximation rectangles shall give a better result; so cross the interval in five steps, using the approximation points 0, 1/5, 2/5, and so on to 1. Fit a box for each step using the right end height of each curve piece, thus √(1D 5), √(2D 5), and so on to √1 = 1. Summing the areas of these rectangles, we get a better approximation for the sought integral, namely

$$\sqrt{\frac{1}{5}}\left(\frac{1}{5}-0\right)+\sqrt{\frac{2}{5}}\left(\frac{2}{5}-\frac{1}{5}\right)+\cdots+\sqrt{\frac{5}{5}}\left(\frac{5}{5}-\frac{4}{5}\right)\approx 0.7497.$$

We are taking a sum of finitely many function values of f, multiplied with the differences of two subsequent approximation points. We can easily see that the approximation is still too large. Using more steps produces a closer approximation, but will never be exact: replacing the

5 subintervals by twelve in the same way, but with the left end height of each piece, we will get an approximate value for the area of 0.6203, which is too small. The key idea is the transition from adding *finitely many* differences of approximation points multiplied by their respective function values to using infinitely many fine, or infinitesimal steps.

As for the *actual calculation of integrals*, the fundamental theorem of calculus, due to Newton and Leibniz, is the fundamental link between the operations of differentiating and integrating. Applied to the square root curve, $f(x) = x^{1/2}$, it says to look at the antiderivative $F(x) = (2/3)x^{3/2}$, and simply take $F(1) - F(0)$, where 0 and 1 are the boundaries of the interval [0,1]. So the *exact* value of the area under the curve is computed formally as

$$\int_0^1 \sqrt{x}\,dx = \int_0^1 x^{1/2}\,dx = F(1) - F(0) = \frac{2}{3}.$$

(This is a case of a general rule, that for $f(x) = x^q$, with $q \neq -1$, the related function, the so-called antiderivative is $F(x) = x^{q+1}/(q + 1)$.)

The notation

$$\int f(x)dx$$

conceives the integral as a weighted sum, denoted by the elongated *s*, of function values, $f(x)$, multiplied by infinitesimal step widths, the so-called *differentials*, denoted by dx. The multiplication sign is usually omitted.

Historically, after the failure of early efforts to rigorously interpret infinitesimals, Riemann formally defined integrals as a limit of weighted sums, so that the dx suggested the limit of a difference (namely, the interval width). Shortcomings of Riemann's dependence on intervals and continuity motivated newer definitions, especially the Lebesgue integral, which is founded on an ability to extend the idea of "measure" in much more flexible ways. Thus the notation

$$\int_A f(x)d\mu$$

refers to a weighted sum in which the function values are partitioned, with μ measuring the weight to be assigned to each value. Here A denotes the region of integration.

Differential geometry, with its "calculus on manifolds", gives the familiar notation yet another interpretation. Now $f(x)$ and dx become a differential form, $\omega = f(x)\ \ dx$, a new differential operator d, known as the exterior derivative is introduced, and the fundamental theorem becomes the more general Stokes' theorem,

$$\int_A d\omega = \int_{\partial A} \omega,$$

from which Green's theorem, the divergence theorem, and the fundamental theorem of calculus follow.

More recently, infinitesimals have reappeared with rigor, through modern innovations such as non-standard analysis. Not only do these methods vindicate the intuitions of the pioneers; they also lead to new mathematics.

Although there are differences between these conceptions of integral, there is considerable overlap. Thus, the area of the surface of the oval swimming pool can be handled as a geometric ellipse, a sum of infinitesimals, a Riemann integral, a Lebesgue integral, or as a manifold with a differential form. The calculated result will be the same for all.

Darboux Sums

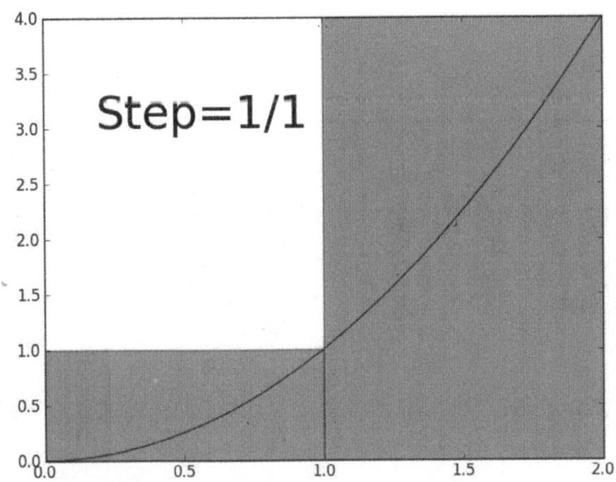

Figure: *Darboux upper sums of the function* $y = x^2$

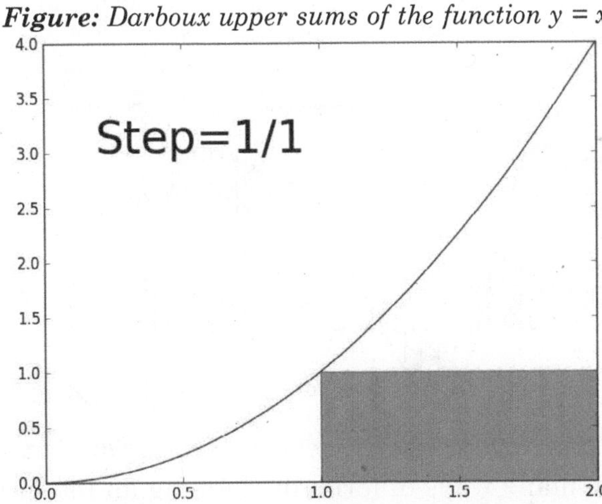

Figure: *Darboux lower sums of the function* $y = x^2$

Formal Definitions

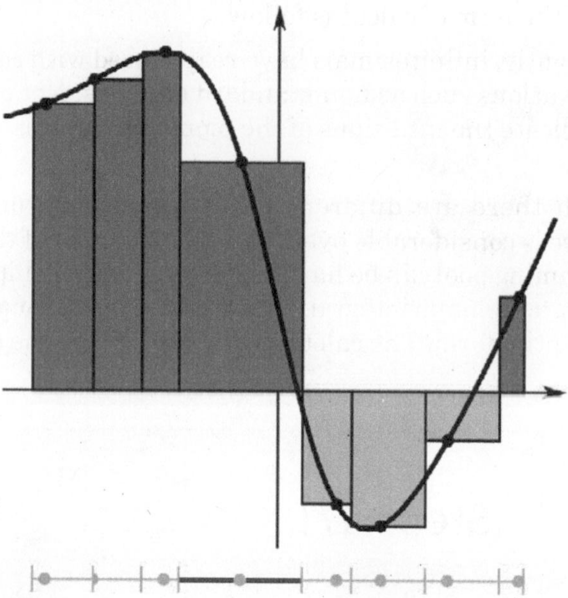

***Figure:** Integral example with irregular partitions*

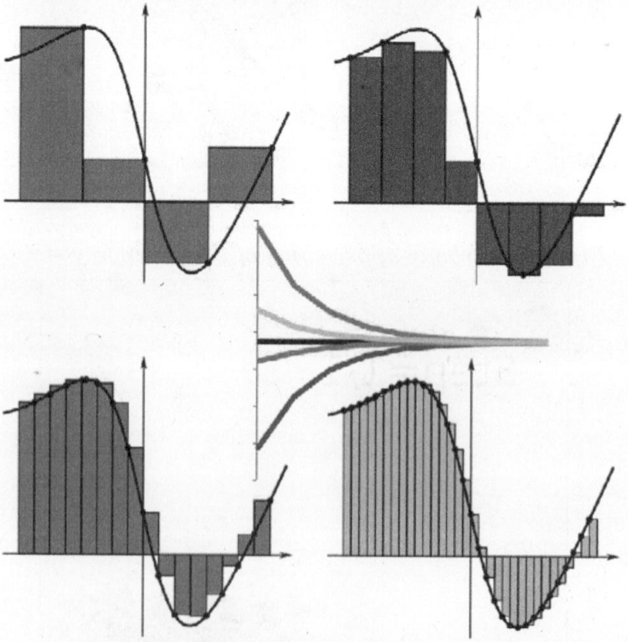

***Figure:** Riemann sums converging*

There are many ways of formally defining an integral, not all of which are equivalent. The differences exist mostly to deal with differing special cases which may not be integrable under other definitions, but also occasionally for pedagogical reasons. The most commonly used definitions of integral are Riemann integrals and Lebesgue integrals.

Riemann Integral

The Riemann integral is defined in terms of Riemann sums of functions with respect to *tagged partitions* of an interval. Let $[a,b]$ be a closed interval of the real line; then a *tagged partition* of $[a,b]$ is a finite sequence

$$a = x_0 \le t_1 \le x_1 \le t_2 \le x_2 \le \cdots \le x_{n-1} \le t_n \le x_n = b.$$

This partitions the interval $[a,b]$ into n sub-intervals $[x_{i-1}, x_i]$ indexed by i, each of which is "tagged" with a distinguished point t_i – $[x_{i-1}, x_i]$. A *Riemann sum* of a function f with respect to such a tagged partition is defined as

$$\sum_{i=1}^{n} f(t_i)\Delta_i;$$

thus each term of the sum is the area of a rectangle with height equal to the function value at the distinguished point of the given sub-interval, and width the same as the sub-interval width. Let $\Delta_i = x_i - x_{i-1}$ be the width of sub-interval i; then the *mesh* of such a tagged partition is the width of the largest sub-interval formed by the partition, $\max_{i=1\ldots n} \Delta_i$. The *Riemann integral* of a function f over the interval $[a,b]$ is equal to S if:

For all $\varepsilon > 0$ there exists $\delta > 0$ such that, for any tagged partition $[a,b]$ with mesh less than δ, we have

$$\left| S - \sum_{i=1}^{n} f(t_i)\Delta_i \right| < \varepsilon.$$

When the chosen tags give the maximum (respectively, minimum) value of each interval, the Riemann sum becomes an upper (respectively, lower) Darboux sum, suggesting the close connection between the Riemann integral and the Darboux integral.

Lebesgue Integral

It is often of interest, both in theory and applications, to be able to pass to the limit under the integral. For instance, a sequence of functions can frequently be constructed that approximate, in a suitable sense, the solution to a problem. Then the integral of the solution function should be the limit of the integrals of the approximations. However, many functions that can be obtained as limits are not Riemann integrable, and so such limit theorems do not hold with the Riemann integral. Therefore it is of great importance to have a definition of the integral that allows a wider class of functions to be integrated (Rudin 1987).

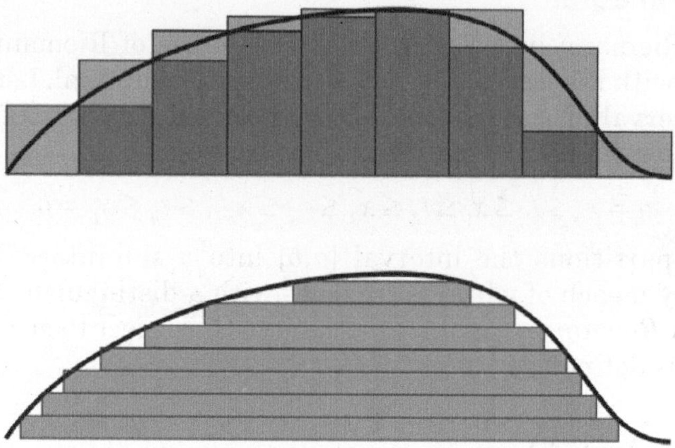

Figure: *Riemann–Darboux's integration (top) and Lebesgue integration (bottom)*

Such an integral is the Lebesgue integral, that exploits the following fact to enlarge the class of integrable functions: if the values of a function are rearranged over the domain, the integral of a function should remain the same. Thus Henri Lebesgue introduced the integral bearing his name, explaining this integral thus in a letter to Paul Montel:

> *I have to pay a certain sum, which I have collected in my pocket. I take the bills and coins out of my pocket and give them to the creditor in the order I find them until I have reached the total sum. This is the Riemann integral. But I can proceed differently. After I have taken all the money out of my pocket I order the bills and coins according to identical values and then I pay the several heaps one after the other to the creditor. This is my integral.*
>
> *Source: (Siegmund-Schultze 2008)*

As Folland (1984, p. 56) puts it, "To compute the Riemann integral of f, one partitions the domain $[a,b]$ into subintervals", while in the Lebesgue integral, "one is in effect partitioning the range of f". The definition of the Lebesgue integral thus begins with a measure, μ. In the simplest case, the Lebesgue measure $\mu(A)$ of an interval $A = [a,b]$ is its width, $b - a$, so that the Lebesgue integral agrees with the (proper)

Riemann integral when both exist. In more complicated cases, the sets being measured can be highly fragmented, with no continuity and no resemblance to intervals.

Using the "partitioning the range of *f*" philosophy, the integral of a non-negative function $f : R \to R$ should be the sum over t of the areas between a thin horizontal strip between $y = t$ and $y = t + dt$. This area is just $\mu\{ x : f(x) > t\}\, dt$. Let $f^*(t) = \mu\{ x : f(x) > t\}$. The Lebesgue integral of f is then defined by (Lieb & Loss 2001)

$$\int f = \int_0^\infty f^*(t)dt$$

where the integral on the right is an ordinary improper Riemann integral (f^* is a strictly decreasing positive function, and therefore has a well-defined improper Riemann integral). For a suitable class of functions (the measurable functions) this defines the Lebesgue integral.

A general measurable function f is Lebesgue integrable if the area between the graph of f and the x-axis is finite:

$$\int_E |f|\, d\mu < +\infty.$$

In that case, the integral is, as in the Riemannian case, the difference between the area above the x-axis and the area below the x-axis:

$$\int_E f\, d\mu = \int_E f^+\, d\mu - \int_E f^-\, d\mu$$

where

$$f^+(x) = \max(\{f(x),0\}) = \begin{cases} f(x), & \text{if } f(x) > 0, \\ 0, & \text{otherwise,} \end{cases}$$

$$f^-(x) = \max(\{-f(x),0\}) = \begin{cases} -f(x), & \text{if } f(x) < 0, \\ 0, & \text{otherwise.} \end{cases}$$

Other Integrals

Although the Riemann and Lebesgue integrals are the most widely used definitions of the integral, a number of others exist, including:

- The Darboux integral which is equivalent to a Riemann integral, meaning that a function is Darboux-integrable if and only if it is Riemann-integrable, and the values of the two integrals, if they exist, are equal. Darboux integrals have the advantage of being simpler to define than Riemann integrals.

- The Riemann–Stieltjes integral, an extension of the Riemann integral.
- The Lebesgue–Stieltjes integral, further developed by Johann Radon, which generalizes the Riemann–Stieltjes and Lebesgue integrals.
- The Daniell integral, which subsumes the Lebesgue integral and Lebesgue–Stieltjes integral without the dependence on measures.
- The Haar integral, used for integration on locally compact topological groups, introduced by Alfréd Haar in 1933.
- The Henstock–Kurzweil integral, variously defined by Arnaud Denjoy, Oskar Perron, and (most elegantly, as the gauge integral) Jaroslav Kurzweil, and developed by Ralph Henstock.
- The Itô integral and Stratonovich integral, which define integration with respect to semimartingales such as Brownian motion.
- The Young integral, which is a kind of Riemann–Stieltjes integral with respect to certain functions of unbounded variation.
- The rough path integral defined for functions equipped with some additional "rough path" structure, generalizing stochastic integration against both semimartingales and processes such as the fractional Brownian motion.

Properties

Linearity

- The collection of Riemann integrable functions on a closed interval $[a, b]$ forms a vector space under the operations of pointwise addition and multiplication by a scalar, and the operation of integration

  $$f \mapsto \int_a^b f(x)\,dx$$

  is a linear functional on this vector space. Thus, firstly, the collection of integrable functions is closed under taking linear combinations; and, secondly, the integral of a linear combination is the linear combination of the integrals,

  $$\int_a^b (\alpha f + \beta g)(x)dx = \alpha \int_a^b f(x)dx + \beta \int_a^b g(x)dx.$$

- Similarly, the set of real-valued Lebesgue integrable functions on a given measure space E with measure μ is closed under taking linear combinations and hence form a vector space, and the Lebesgue integral

$$f \mapsto \int_E f \, d\mu$$

is a linear functional on this vector space, so that

$$\int_E (\alpha f + \beta g) d\mu = \alpha \int_E f \, d\mu + \beta \int_E g \, d\mu.$$

- More generally, consider the vector space of all measurable functions on a measure space (E,μ), taking values in a locally compact complete topological vector space V over a locally compact topological field K, $f : E \to V$. Then one may define an abstract integration map assigning to each function f an element of V or the symbol ∞,

  $$f \mapsto \int_E f \, d\mu,$$

  that is compatible with linear combinations. In this situation the linearity holds for the subspace of functions whose integral is an element of V (i.e. "finite"). The most important special cases arise when K is R, C, or a finite extension of the field Q_p of p-adic numbers, and V is a finite-dimensional vector space over K, and when K=C and V is a complex Hilbert space.

Linearity, together with some natural continuity properties and normalisation for a certain class of "simple" functions, may be used to give an alternative definition of the integral. This is the approach of Daniell for the case of real-valued functions on a set X, generalized by Nicolas Bourbaki to functions with values in a locally compact topological vector space.

Inequalities

A number of general inequalities hold for Riemann-integrable functions defined on a closed and bounded interval $[a, b]$ and can be generalized to other notions of integral (Lebesgue and Daniell).

- *Upper and lower bounds.* An integrable function f on $[a, b]$, is necessarily bounded on that interval. Thus there are real numbers m and M so that $m \le f(x) \le M$ for all x in $[a, b]$. Since the lower and upper sums of f over $[a, b]$ are therefore bounded by, respectively, $m(b - a)$ and $M(b - a)$, it follows that

  $$m(b-a) \le \int_a^b f(x)dx \le M(b-a).$$

- *Inequalities between functions.* If $f(x) \le g(x)$ for each x in $[a, b]$ then each of the upper and lower sums of f is bounded above by the upper and lower sums, respectively, of g. Thus

$$\int_a^b f(x)dx \le \int_a^b g(x)dx.$$

This is a generalization of the above inequalities, as $M(b-a)$ is the integral of the constant function with value M over $[a, b]$.

In addition, if the inequality between functions is strict, then the inequality between integrals is also strict. That is, if $f(x) < g(x)$ for each x in $[a, b]$, then

$$\int_a^b f(x)dx < \int_a^b g(x)dx.$$

- *Subintervals.* If $[c, d]$ is a subinterval of $[a, b]$ and $f(x)$ is non-negative for all x, then

$$\int_c^d f(x)dx \le \int_a^b f(x)dx.$$

- *Products and absolute values of functions.* If f and g are two functions then we may consider their pointwise products and powers, and absolute values:

$$(fg)(x) = f(x)g(x),\ f^2(x) = (f(x))^2,\ |f|(x) = |f(x)|.$$

If f is Riemann-integrable on $[a, b]$ then the same is true for $|f|$, and

$$\left|\int_a^b f(x)dx\right| \le \int_a^b |f(x)|dx.$$

Moreover, if f and g are both Riemann-integrable then fg is also Riemann-integrable, and

$$\left(\int_a^b (fg)(x)dx\right)^2 \le \left(\int_a^b f(x)^2 dx\right)\left(\int_a^b g(x)^2 dx\right).$$

This inequality, known as the Cauchy–Schwarz inequality, plays a prominent role in Hilbert space theory, where the left hand side is interpreted as the inner product of two square-integrable functions f and g on the interval $[a, b]$.

- *Hölder's inequality.* Suppose that p and q are two real numbers, $1 \le p, q \le \infty$ with $1/p + 1/q = 1$, and f and g are two Riemann-integrable functions. Then the functions $|f|^p$ and $|g|^q$ are also integrable and the following Hölder's inequality holds:

$$\left|\int f(x)g(x)dx\right| \le \left(\int |f(x)|^p dx\right)^{1/p} \left(\int |g(x)|^q dx\right)^{1/q}.$$

For $p = q = 2$, Hölder's inequality becomes the Cauchy–Schwarz inequality.

- *Minkowski inequality*. Suppose that $p \leq 1$ is a real number and f and g are Riemann-integrable functions. Then $|f|^p$, $|g|^p$ and $|f + g|^p$ are also Riemann integrable and the following Minkowski inequality holds:

$$\left(\int |f(x)+g(x)|^p\, dx\right)^{1/p} \leq \left(\int |f(x)|^p\, dx\right)^{1/p} + \left(\int |g(x)|^p\, dx\right)^{1/p}.$$

An analogue of this inequality for Lebesgue integral is used in construction of Lp spaces.

Conventions

In this section f is a real-valued Riemann-integrable function. The integral

$$\int_a^b f(x)dx$$

over an interval $[a, b]$ is defined if $a < b$. This means that the upper and lower sums of the function f are evaluated on a partition $a = x_0 \leq x_1 \leq \ldots \leq x_n = b$ whose values x_i are increasing. Geometrically, this signifies that integration takes place "left to right", evaluating f within intervals $[x_i, x_{i+1}]$ where an interval with a higher index lies to the right of one with a lower index. The values a and b, the end-points of the interval, are called the limits of integration of f. Integrals can also be defined if $a > b$:

- *Reversing limits of integration.* If $a > b$ then define

$$\int_a^b f(x)dx = -\int_b^a f(x)dx.$$

This, with $a = b$, implies:

- *Integrals over intervals of length zero.* If a is a real number then

$$\int_a^a f(x)dx = 0.$$

The first convention is necessary in consideration of taking integrals over subintervals of $[a, b]$; the second says that an integral taken over a degenerate interval, or a point, should be zero. One reason for the first convention is that the integrability of f on an interval $[a, b]$ implies that f is integrable on any subinterval $[c, d]$, but in particular integrals have the property that:

- *Additivity of integration on intervals.* If c is any element of $[a, b]$, then

$$\int_a^b f(x)dx = \int_a^c f(x)dx + \int_c^b f(x)dx.$$

With the first convention the resulting relation

$$\begin{aligned}\int_a^c f(x)dx &= \int_a^b f(x)dx - \int_c^b f(x)dx \\ &= \int_a^b f(x)dx + \int_b^c f(x)dx\end{aligned}$$

is then well-defined for any cyclic permutation of *a*, *b*, and *c*.

Instead of viewing the above as conventions, one can also adopt the point of view that integration is performed of differential forms on oriented manifolds only. If *M* is such an oriented *m*-dimensional manifold, and *M'* is the same manifold with opposed orientation and ω is an *m*-form, then one has:

$$\int_M \omega = -\int_{M'} \omega.$$

These conventions correspond to interpreting the integrand as a differential form, integrated over a chain. In measure theory, by contrast, one interprets the integrand as a function *f* with respect to a measure μ, and integrates over a subset *A*, without any notion of orientation; one writes $\int_A f\,d\mu = \int_{[a,b]} f\,d\mu$ to indicate integration over a subset *A*. This is a minor distinction in one dimension, but becomes subtler on higher-dimensional manifolds.

Fundamental Theorem of Calculus

The *fundamental theorem of calculus* is the statement that differentiation and integration are inverse operations: if a continuous function is first integrated and then differentiated, the original function is retrieved. An important consequence, sometimes called the *second fundamental theorem of calculus*, allows one to compute integrals by using an antiderivative of the function to be integrated.

Statements of Theorems

Fundamental Theorem of Calculus: Let *f* be a continuous real-valued function defined on a closed interval [*a*, *b*]. Let *F* be the function defined, for all *x* in [*a*, *b*], by

$$F(x) = \int_a^x f(t)dt.$$

Then, *F* is continuous on [*a*, *b*], differentiable on the open interval (*a*, *b*), and

$$F'(x) = f(x)$$

for all *x* in (*a*, *b*).

Second Fundamental Theorem of Calculus

Let f be a real-valued function defined on a closed interval $[a, b]$ that admits an antiderivative F on $[a, b]$. That is, f and F are functions such that for all x in $[a, b]$,

$$f(x) = F'(x).$$

If f is integrable on $[a, b]$ then

$$\int_a^b f(x)dx = F(b) - F(a).$$

Extensions

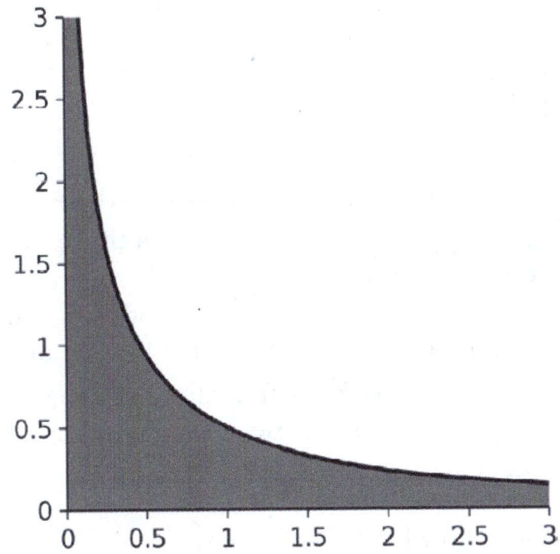

Figure: *The improper integral* $\int_0^\infty \frac{dx}{(x+1)\sqrt{x}} = \pi$ *has unbounded intervals for both domain and range.*

A "proper" Riemann integral assumes the integrand is defined and finite on a closed and bounded interval, bracketed by the limits of integration. An improper integral occurs when one or more of these conditions is not satisfied. In some cases such integrals may be defined by considering the limit of a sequence of proper Riemann integrals on progressively larger intervals.

If the interval is unbounded, for instance at its upper end, then the improper integral is the limit as that endpoint goes to infinity.

$$\int_a^\infty f(x)dx = \lim_{b \to \infty} \int_a^b f(x)dx$$

If the integrand is only defined or finite on a half-open interval, for instance $(a, b]$, then again a limit may provide a finite result.

$$\int_a^b f(x)dx = \lim_{\epsilon \to 0} \int_{a+\epsilon}^b f(x)dx$$

That is, the improper integral is the limit of proper integrals as one endpoint of the interval of integration approaches either a specified real number, or ∞, or $-\infty$. In more complicated cases, limits are required at both endpoints, or at interior points.

Consider, for example, the function $1/((x+1)\sqrt{x})$ integrated from 0 to ∞ (shown right). At the lower bound, as x goes to 0 the function goes to ∞, and the upper bound is itself ∞, though the function goes to 0. Thus this is a doubly improper integral. Integrated, say, from 1 to 3, an ordinary Riemann sum suffices to produce a result of $\pi/6$. To integrate from 1 to ∞, a Riemann sum is not possible.

However, any finite upper bound, say t (with $t > 1$), gives a well-defined result, $2\arctan(\sqrt{t}) - \pi/2$. This has a finite limit as t goes to infinity, namely $\pi/2$. Similarly, the integral from 1/3 to 1 allows a Riemann sum as well, coincidentally again producing $\pi/6$. Replacing 1/3 by an arbitrary positive value s (with $s < 1$) is equally safe, giving $\pi/2 - 2\arctan(\sqrt{s})$. This, too, has a finite limit as s goes to zero, namely $\pi/2$. Combining the limits of the two fragments, the result of this improper integral is

$$\begin{aligned}\int_0^\infty \frac{dx}{(x+1)\sqrt{x}} &= \lim_{s\to 0}\int_s^1 \frac{dx}{(x+1)\sqrt{x}} + \lim_{t\to\infty}\int_1^t \frac{dx}{(x+1)\sqrt{x}} \\ &= \lim_{s\to 0}\left(\frac{\pi}{2} - 2\arctan\sqrt{s}\right) + \lim_{t\to\infty}\left(2\arctan\sqrt{t} - \frac{\pi}{2}\right) \\ &= \frac{\pi}{2} + \left(\pi - \frac{\pi}{2}\right) \\ &= \frac{\pi}{2} + \frac{\pi}{2} \\ &= \pi.\end{aligned}$$

This process does not guarantee success; a limit might fail to exist, or might be unbounded. For example, over the bounded interval from 0 to 1 the integral of $1/x$ does not converge; and over the unbounded interval from 1 to ∞ the integral of $1/\sqrt{x}$ does not converge.

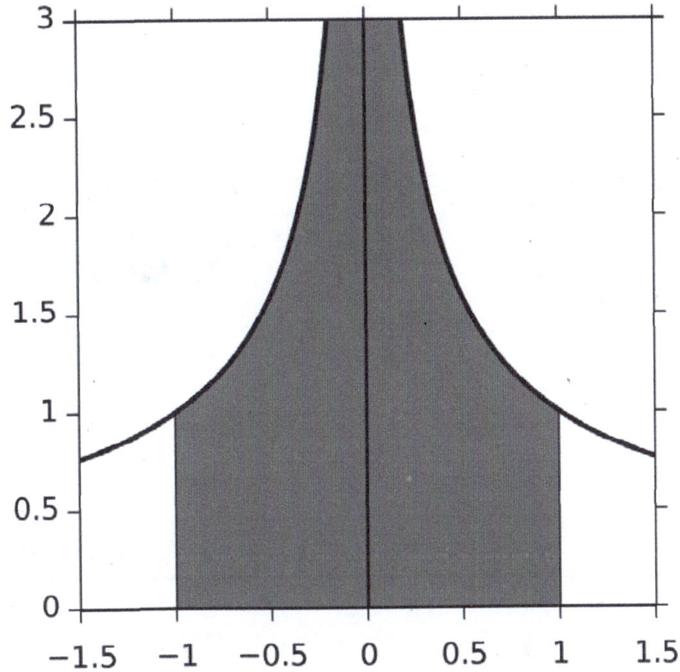

Figure: *The improper integral* $\int_{-1}^{1}\frac{dx}{\sqrt[3]{x^2}}=6$ *is unbounded internally, but both left and right limits exist.*

It might also happen that an integrand is unbounded at an interior point, in which case the integral must be split at that point. For the integral as a whole to converge, the limit integrals on both sides must exist and must be bounded. For example:

$$\begin{aligned}\int_{-1}^{1}\frac{dx}{\sqrt[3]{x^2}} &= \lim_{s\to 0}\int_{-1}^{-s}\frac{dx}{\sqrt[3]{x^2}}+\lim_{t\to 0}\int_{t}^{1}\frac{dx}{\sqrt[3]{x^2}}\\ &= \lim_{s\to 0}3(1-\sqrt[3]{s})+\lim_{t\to 0}3(1-\sqrt[3]{t})\\ &= 3+3\\ &= 6.\end{aligned}$$

But the similar integral

$$\int_{-1}^{1}\frac{dx}{x}$$

cannot be assigned a value in this way, as the integrals above and below zero do not independently converge.

Multiple Integration

Figure: Double integral as volume under a surface.

Integrals can be taken over regions other than intervals. In general, an integral over a set E of a function f is written:

$$\int_E f(x)dx.$$

Here x need not be a real number, but can be another suitable quantity, for instance, a vector in R^3. Fubini's theorem shows that such integrals can be rewritten as an iterated integral. In other words, the integral can be calculated by integrating one coordinate at a time.

Just as the definite integral of a positive function of one variable represents the area of the region between the graph of the function and the x-axis, the *double integral* of a positive function of two variables represents the volume of the region between the surface defined by the function and the plane which contains its domain. (The same volume can be obtained via the *triple integral* — the integral of a function in three variables — of the constant function $f(x, y, z) = 1$ over the above mentioned region between the surface and the plane.) If the number of variables is higher, then the integral represents a hypervolume, a volume of a solid of more than three dimensions that cannot be graphed.

For example, the volume of the cuboid of sides 4 × 6 × 5 may be obtained in two ways:

- By the double integral

  $$\iint_D 5 \, dxdy$$

  of the function $f(x, y) = 5$ calculated in the region D in the xy-plane which is the base of the cuboid. For example, if a rectangular base of such a cuboid is given via the xy inequalities $3 \leq x \leq 7$, $4 \leq y \leq 10$, our above double integral now reads

  $$\int_4^{10}\left[\int_3^7 5 \, dx\right]dy.$$

  From here, integration is conducted with respect to either x or y first; in this example, integration is first done with respect to x as the interval corresponding to x is the inner integral. Once the first integration is completed via the $F(b) - F(a)$ method or otherwise, the result is again integrated with respect to the other variable. The result will equate to the volume under the surface.

- By the triple integral

  $$\iiint_{\text{cuboid}} 1dxdydz$$

  of the constant function 1 calculated on the cuboid itself.

Line Integrals

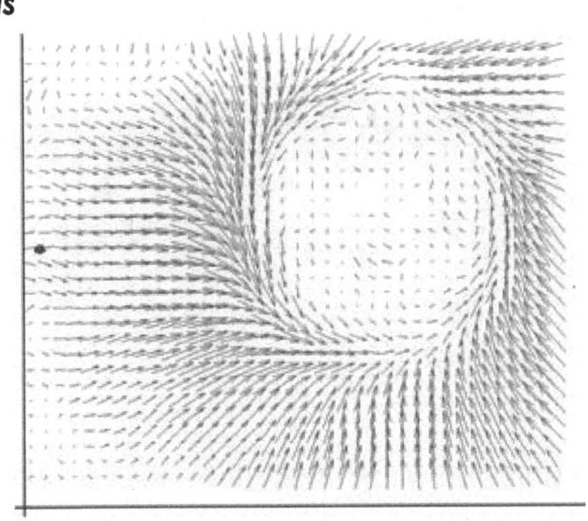

Figure: *A line integral sums together elements along a curve.*

The concept of an integral can be extended to more general domains of integration, such as curved lines and surfaces. Such integrals are known as line integrals and surface integrals respectively. These have important applications in physics, as when dealing with vector fields.

A *line integral* (sometimes called a *path integral*) is an integral where the function to be integrated is evaluated along a curve. Various different line integrals are in use. In the case of a closed curve it is also called a *contour integral.*

The function to be integrated may be a scalar field or a vector field. The value of the line integral is the sum of values of the field at all points on the curve, weighted by some scalar function on the curve (commonly arc length or, for a vector field, the scalar product of the vector field with a differential vector in the curve). This weighting distinguishes the line integral from simpler integrals defined on intervals. Many simple formulas in physics have natural continuous analogs in terms of line integrals; for example, the fact that work is equal to force, F, multiplied by displacement, s, may be expressed (in terms of vector quantities) as:

$$W = \mathbf{F} \cdot \mathbf{s}.$$

For an object moving along a path C in a vector field F such as an electric field or gravitational field, the total work done by the field on the object is obtained by summing up the differential work done in moving from s to s + ds. This gives the line integral

$$W = \int_C \mathbf{F} \cdot d\mathbf{s}.$$

Surface Integrals

Figure: *The definition of surface integral relies on splitting the surface into small surface elements.*

A *surface integral* is a definite integral taken over a surface (which may be a curved set in space); it can be thought of as the double integral analog of the line integral. The function to be integrated may be a scalar field or a vector field. The value of the surface integral is the sum of the field at all points on the surface. This can be achieved by splitting the surface into surface elements, which provide the partitioning for Riemann sums.

For an example of applications of surface integrals, consider a vector field v on a surface S; that is, for each point x in S, v(x) is a vector. Imagine that we have a fluid flowing through S, such that v(x) determines the velocity of the fluid at x. The flux is defined as the quantity of fluid flowing through S in unit amount of time. To find the flux, we need to take the dot product of v with the unit surface normal to S at each point, which will give us a scalar field, which we integrate over the surface:

$$\int_S \mathbf{v}\cdot d\mathbf{S}.$$

The fluid flux in this example may be from a physical fluid such as water or air, or from electrical or magnetic flux. Thus surface integrals have applications in physics, particularly with the classical theory of electromagnetism.

Integrals of Differential Forms

A differential form is a mathematical concept in the fields of multivariable calculus, differential topology and tensors. The modern notation for the differential form, as well as the idea of the differential forms as being the wedge products of exterior derivatives forming an exterior algebra, was introduced by Élie Cartan.

We initially work in an open set in R^n. A 0-form is defined to be a smooth function f. When we integrate a function f over an m-dimensional subspace S of R^n, we write it as

$$\int_S f\, dx^1 \dots dx^m.$$

(The superscripts are indices, not exponents.) We can consider dx^1 through dx^n to be formal objects themselves, rather than tags appended to make integrals look like Riemann sums. Alternatively, we can view them as covectors, and thus a measure of "density" (hence integrable in a general sense). We call the dx^1, ..., dx^n *basic* 1-forms.

We define the wedge product, "$\wedge$", a bilinear "multiplication" operator on these elements, with the *alternating* property that

$$dx^a \wedge dx^a = 0$$

for all indices a. Alternation along with linearity and associativity implies $dx^b \wedge dx^a = -dx^a \wedge dx^b$. This also ensures that the result of the wedge product has an orientation.

We define the set of all these products to be *basic 2-forms*, and similarly we define the set of products of the form $dx^a \wedge dx^b \wedge dx^c$ to be *basic 3-forms*. A general k-form is then a weighted sum of basic k-forms, where the weights are the smooth functions f. Together these form a vector space with basic k-forms as the basis vectors, and 0-forms (smooth functions) as the field of scalars. The wedge product then extends to k-forms in the natural way. Over R^n at most n covectors can be linearly independent, thus a k-form with $k > n$ will always be zero, by the alternating property.

In addition to the wedge product, there is also the exterior derivative operator d. This operator maps k-forms to $(k+1)$-forms. For a k-form $\omega = f\, dx^a$ over R^n, we define the action of d by:

$$d\omega = \sum_{i=1}^{n} \frac{\partial f}{\partial x_i} dx^i \wedge dx^a.$$

with extension to general k-forms occurring linearly.

This more general approach allows for a more natural coordinate-free approach to integration on manifolds. It also allows for a natural generalisation of the fundamental theorem of calculus, called Stokes' theorem, which we may state as

$$\int_{\Omega} d\omega = \int_{\partial\Omega} \omega$$

where ω is a general k-form, and $\partial\Omega$ denotes the boundary of the region Ω. Thus, in the case that Ω is a 0-form and Ω is a closed interval of the real line, this reduces to the fundamental theorem of calculus. In the case that ω is a 1-form and Ω is a two-dimensional region in the plane, the theorem reduces to Green's theorem. Similarly, using 2-forms, and 3-forms and Hodge duality, we can arrive at Stokes' theorem and the divergence theorem. In this way we can see that differential forms provide a powerful unifying view of integration.

Summations

The discrete equivalent of integration is summation. Summations and integrals can be put on the same foundations using the theory of Lebesgue integrals or time scale calculus.

Computation

Analytical: The most basic technique for computing definite integrals of one real variable is based on the fundamental theorem of calculus. Let $f(x)$ be the function of x to be integrated over a given interval $[a, b]$. Then, find an antiderivative of f; that is, a function F

such that $F' = f$ on the interval. Provided the integrand and integral have no singularities on the path of integration, by the fundamental theorem of calculus,

$$\int_a^b f(x)dx = F(b) - F(a).$$

The integral is not actually the antiderivative, but the fundamental theorem provides a way to use antiderivatives to evaluate definite integrals.

The most difficult step is usually to find the antiderivative of f. It is rarely possible to glance at a function and write down its antiderivative. More often, it is necessary to use one of the many techniques that have been developed to evaluate integrals. Most of these techniques rewrite one integral as a different one which is hopefully more tractable. Techniques include:

- Integration by substitution
- Integration by parts
- Inverse function integration
- Changing the order of integration
- Integration by trigonometric substitution
- Tangent half-angle substitution
- Integration by partial fractions
- Integration by reduction formulae
- Integration using parametric derivatives
- Integration using Euler's formula
- Euler substitution
- Differentiation under the integral sign
- Contour integration

Alternative methods exist to compute more complex integrals. Many nonelementary integrals can be expanded in a Taylor series and integrated term by term. Occasionally, the resulting infinite series can be summed analytically. The method of convolution using Meijer G-functions can also be used, assuming that the integrand can be written as a product of Meijer G-functions. There are also many less common ways of calculating definite integrals; for instance, Parseval's identity can be used to transform an integral over a rectangular region into an infinite sum. Occasionally, an integral can be evaluated by a trick.

Computations of volumes of solids of revolution can usually be done with disk integration or shell integration.

Specific results which have been worked out by various techniques are collected in the list of integrals.

Symbolic

Many problems in mathematics, physics, and engineering involve integration where an explicit formula for the integral is desired. Extensive tables of integrals have been compiled and published over the years for this purpose. With the spread of computers, many professionals, educators, and students have turned to computer algebra systems that are specifically designed to perform difficult or tedious tasks, including integration. Symbolic integration has been one of the motivations for the development of the first such systems, like Macsyma.

A major mathematical difficulty in symbolic integration is that in many cases, a closed formula for the antiderivative of a rather simple-looking function does not exist. For instance, it is known that the antiderivatives of the functions $\exp(x^2)$, x^x and $(\sin x)/x$ cannot be expressed in the closed form involving only rational and exponential functions, logarithm, trigonometric and inverse trigonometric functions, and the operations of multiplication and composition; in other words, none of the three given functions is integrable in elementary functions, which are the functions which may be built from rational functions, roots of a polynomial, logarithm, and exponential functions. The Risch algorithm provides a general criterion to determine whether the antiderivative of an elementary function is elementary, and, if it is, to compute it. Unfortunately, it turns out that functions with closed expressions of antiderivatives are the exception rather than the rule. Consequently, computerized algebra systems have no hope of being able to find an antiderivative for a randomly constructed elementary function. On the positive side, if the 'building blocks' for antiderivatives are fixed in advance, it may be still be possible to decide whether the antiderivative of a given function can be expressed using these blocks and operations of multiplication and composition, and to find the symbolic answer whenever it exists. The Risch algorithm, implemented in Mathematica and other computer algebra systems, does just that for functions and antiderivatives built from rational functions, radicals, logarithm, and exponential functions.

Some special integrands occur often enough to warrant special study. In particular, it may be useful to have, in the set of antiderivatives, the special functions of physics (like the Legendre functions, the hypergeometric function, the Gamma function, the Incomplete Gamma function and so on). Extending the Risch's algorithm to include such functions is possible but challenging and has been an active research subject.

More recently a new approach has emerged, using D-finite function, which are the solutions of linear differential equations with polynomial coefficients. Most of the elementary and special functions are *D*-finite and the integral of a *D*-finite function is also a *D*-finite function. This provide an algorithm to express the antiderivative of a *D*-finite function as the solution of a differential equation.

This theory allows also to compute a definite integrals of a *D*-function as the sum of a series given by the first coefficients and an algorithm to compute any coefficient.

Numerical

The integrals encountered in a basic calculus course are deliberately chosen for simplicity; those found in real applications are not always so accommodating. Some integrals cannot be found exactly, some require special functions which themselves are a challenge to compute, and others are so complex that finding the exact answer is too slow. This motivates the study and application of numerical methods for approximating integrals, which today use floating-point arithmetic on digital electronic computers. Many of the ideas arose much earlier, for hand calculations; but the speed of general-purpose computers like the ENIAC created a need for improvements.

The goals of numerical integration are accuracy, reliability, efficiency, and generality. Sophisticated methods can vastly outperform a naive method by all four measures (Dahlquist & Björck 2008; Kahaner, Moler & Nash 1989; Stoer & Bulirsch 2002). Consider, for example, the integral

$$\int_{-2}^{2} \frac{1}{5}\left(\frac{1}{100}(322+3x(98+x(37+x)))-24\frac{x}{1+x^2}\right)dx$$

which has the exact answer 94/25 = 3.76. (In ordinary practice the answer is not known in advance, so an important task — not explored here — is to decide when an approximation is good enough.) A "calculus book" approach divides the integration range into, say, 16 equal pieces, and computes function values.

Spaced function values									
x	–2.00	–1.50	–1.00	–0.50	0.00	0.50	1.00	1.50	2.00
$f(x)$	2.22800	2.45663	2.67200	2.32475	0.64400	–0.92575	–0.94000	–0.16963	0.83600
x	–1.75	–1.25	–0.75	–0.25	0.25	0.75	1.25	1.75	
$f(x)$	2.33041	2.58562	2.62934	1.64019	–0.32444	–1.09159	–0.60387	0.31734	

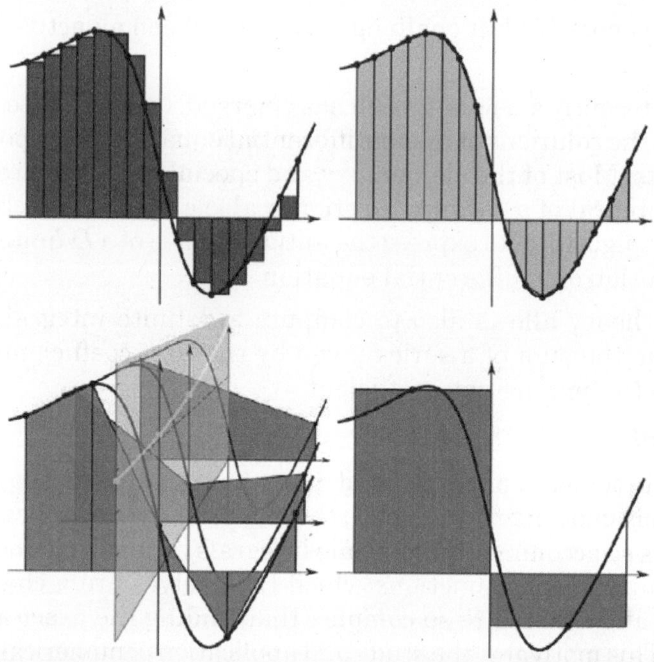

Figure: *Numerical quadrature methods: Rectangle, Trapezoid, Romberg, Gauss*

Using the left end of each piece, the rectangle method sums 16 function values and multiplies by the step width, h, here 0.25, to get an approximate value of 3.94325 for the integral. The accuracy is not impressive, but calculus formally uses pieces of infinitesimal width, so initially this may seem little cause for concern. Indeed, repeatedly doubling the number of steps eventually produces an approximation of 3.76001. However, 2^{18} pieces are required, a great computational expense for such little accuracy; and a reach for greater accuracy can force steps so small that arithmetic precision becomes an obstacle.

A better approach replaces the horizontal tops of the rectangles with slanted tops touching the function at the ends of each piece. This trapezium rule is almost as easy to calculate; it sums all 17 function values, but weights the first and last by one half, and again multiplies by the step width. This immediately improves the approximation to 3.76925, which is noticeably more accurate. Furthermore, only 2^{10} pieces are needed to achieve 3.76000, substantially less computation than the rectangle method for comparable accuracy.

Romberg's method builds on the trapezoid method to great effect. First, the step lengths are halved incrementally, giving trapezoid approximations denoted by $T(h_0)$, $T(h_1)$, and so on, where h_{k+1} is half of h_k. For each new step size, only half the new function values need to be computed; the others carry over from the previous size (as shown in the table above). But the really powerful idea is to interpolate a polynomial through the approximations, and extrapolate to $T(0)$. With this method a numerically *exact* answer here requires only four pieces (five function values)! The Lagrange polynomial interpolating $\{h_k, T(h_k)\}_{k=0...2}$ = {(4.00,6.128), (2.00,4.352), (1.00,3.908)} is $3.76 + 0.148h^2$, producing the extrapolated value 3.76 at $h = 0$.

Gaussian quadrature often requires noticeably less work for superior accuracy. In this example, it can compute the function values at just two x positions, $\pm 2/\sqrt{3}$, then double each value and sum to get the numerically exact answer. The explanation for this dramatic success lies in error analysis, and a little luck. An n-point Gaussian method is exact for polynomials of degree up to $2n$–1. The function in this example is a degree 3 polynomial, plus a term that cancels because the chosen endpoints are symmetric around zero. (Cancellation also benefits the Romberg method.) Shifting the range left a little, so the integral is from –2.25 to 1.75, removes the symmetry. Nevertheless, the trapezoid method is rather slow, the polynomial interpolation method of Romberg is acceptable, and the Gaussian method requires the least work — if the number of points is known in advance. As well, rational interpolation can use the same trapezoid evaluations as the Romberg method to greater effect.

Quadrature method cost comparison				
Method	Trapezoid	Romberg	Rational	Gauss
Points	1048577	257	129	36
Rel. Err.	-5.3×10^{-13}	-6.3×10^{-15}	8.8×10^{-15}	3.1×10^{-15}
Value	$\int_{-2.25}^{1.75} f(x)dx = 4.16390190065858970 75\ldots$			

In practice, each method must use extra evaluations to ensure an error bound on an unknown function; this tends to offset some of the advantage of the pure Gaussian method, and motivates the popular Gauss–Kronrod quadrature formulae. Symmetry can still be exploited by splitting this integral into two ranges, from –2.25 to –1.75 (no symmetry), and from –1.75 to 1.75 (symmetry). More broadly, adaptive quadrature partitions a range into pieces based on function properties, so that data points are concentrated where they are needed most.

Simpson's rule, named for Thomas Simpson (1710–1761), uses a parabolic curve to approximate integrals. In many cases, it is more accurate than the trapezoidal rule and others. The rule states that

$$\int_a^b f(x)dx \approx \frac{b-a}{6}\left[f(a)+4f\left(\frac{a+b}{2}\right)+f(b)\right],$$

with an error of

$$\left| -\frac{(b-a)^5}{2880} f^{(4)}(\xi) \right|.$$

The computation of higher-dimensional integrals (for example, volume calculations) makes important use of such alternatives as Monte Carlo integration. A calculus text is no substitute for numerical analysis, but the reverse is also true. Even the best adaptive numerical code sometimes requires a user to help with the more demanding integrals. For example, improper integrals may require a change of variable or methods that can avoid infinite function values, and known properties like symmetry and periodicity may provide critical leverage.

Mechanical

The area of an arbitrary two-dimensional shape can be determined using a measuring instrument called planimeter. The volume of irregular objects can be measured with precision by the fluid displaced as the object is submerged.

Geometrical

Area can be found via geometrical compass-and-straightedge constructions of an equivalent square, e.g., squaring the circle.

Some Important Definite Integrals

Mathematicians have used definite integrals as a tool to define identities. Among these identities is the definition of the Euler–Mascheroni constant:

$$\gamma = \int_1^\infty \left(\frac{1}{x} - \frac{1}{x} \right) dx,$$

the Gamma function:

$$\Gamma(z) = \int_0^\infty e^{-t} t^{z-1} dt,$$

the Fourier transform which is widely used in physics:

$$F(\xi) = \int_{-\infty}^\infty f(x) e^{-2\pi i x \xi} dx,$$

Laplace transform which is widely used in engineering:

$$F(s) = \int_0^\infty f(t) e^{-st} dt,$$

and the Gaussian Integral fundamental to the Normal Distribution used in probability and statistics:

$$\int_{-\infty}^\infty e^{-x^2} dx = \sqrt{\pi}.$$

Chapter 6

Statistical Description of Data

Descriptive statistics is the discipline of quantitatively describing the main features of a collection of information, or the quantitative description itself. Descriptive statistics are distinguished from inferential statistics (or inductive statistics), in that descriptive statistics aim to summarize a sample, rather than use the data to learn about the population that the sample of data is thought to represent. This generally means that descriptive statistics, unlike inferential statistics, are not developed on the basis of probability theory. Even when a data analysis draws its main conclusions using inferential statistics, descriptive statistics are generally also presented. For example in a paper reporting on a study involving human subjects, there typically appears a table giving the overall sample size, sample sizes in important subgroups (e.g., for each treatment or exposure group), and demographic or clinical characteristics such as the average age, the proportion of subjects of each sex, and the proportion of subjects with related comorbidities.

Some measures that are commonly used to describe a data set are measures of central tendency and measures of variability or dispersion. Measures of central tendency include the mean, median and mode, while measures of variability include the standard deviation (or variance), the minimum and maximum values of the variables, kurtosis and skewness.

Use in Statistical Analysis

Descriptive statistics provides simple summaries about the sample and about the observations that have been made. Such summaries may be either quantitative, i.e. summary statistics, or visual, i.e. simple-to-understand graphs. These summaries may either form the basis of the initial description of the data as part of a more extensive statistical analysis, or they may be sufficient in and of themselves for a particular investigation.

For example, the shooting percentage in basketball is a descriptive statistic that summarizes the performance of a player or a team. This number is the number of shots made divided by the number of shots taken. For example, a player who shoots 33% is making approximately one shot in every three. The percentage summarizes or describes multiple discrete events. Consider also the grade point average. This single number describes the general performance of a student across the range of their course experiences.

The use of descriptive and summary statistics has an extensive history and, indeed, the simple tabulation of populations and of economic data was the first way the topic of statistics appeared. More recently, a collection of summarisation techniques has been formulated under the heading of exploratory data analysis: an example of such a technique is the box plot.

In the business world, descriptive statistics provides a useful summary of many types of data. For example, investors and brokers may use a historical account of return behaviour by performing empirical and analytical analyses on their investments in order to make better investing decisions in the future.

Univariate Analysis

Univariate analysis involves describing the distribution of a single variable, including its central tendency (including the mean, median, and mode) and dispersion (including the range and quantiles of the data-set, and measures of spread such as the variance and standard deviation). The shape of the distribution may also be described via indices such as skewness and kurtosis. Characteristics of a variable's distribution may also be depicted in graphical or tabular format, including histograms and stem-and-leaf display.

Bivariate Analysis

When a sample consists of more than one variable, descriptive statistics may be used to describe the relationship between pairs of variables. In this case, descriptive statistics include:

- Cross-tabulations and contingency tables
- Graphical representation via scatterplots

- Quantitative measures of dependence
- Descriptions of conditional distributions

The main reason for differentiating univariate and bivariate analysis is that bivariate analysis is not only simple descriptive analysis, but also it describes the relationship between two different variables. Quantitative measures of dependence include correlation (such as Pearson's r when both variables are continuous, or Spearman's rho if one or both are not) and covariance (which reflects the scale variables are measured on). The slope, in regression analysis, also reflects the relationship between variables. The unstandardised slope indicates the unit change in the criterion variable for a one unit change in the predictor. The standardised slope indicates this change in standardised (z-score) units. Highly skewed data are often transformed by taking logarithms. Use of logarithms makes graphs more symmetrical and look more similar to the normal distribution, making them easier to interpret intuitively.

Classification and Tabulation

In any statistical investigation, the collection of the numerical data is the first and the most important matter to be attended. Often a person investigating, will have to collect the data from the actual field of inquiry. For this he may issue suitable questionnaires to get necessary information or he may take actual interviews; personal interviews are more effective than questionnaires, which may not evoke an adequate response. Another method of collecting data may be available in publications of Government bodies or other public or private organisations.

Sometimes the data may be available in publications of Government bodies or other public or private organisations. Such data, however, is often so numerous that one's mind can hardly comprehend its significance in the form that it is shown. Therefore it becomes, very necessary to tabulate and summarize the data to an easily manageable form. In doing so we may overlook its details. But this is not a serious loss because Statistics is not interested in an individual but in the properties of aggregates. For a layman, presentation of the raw data in the form of tables or diagrams is always more effective.

Tabulation

It is the process of condensation of the data for convenience, in statistical processing, presentation and interpretation of the information.

A good table is one which has the following requirements :

1. It should present the data clearly, highlighting important details.
2. It should save space but attractively designed.
3. The table number and title of the table should be given.+
4. Row and column headings must explain the figures therein.
5. Averages or percentages should be close to the data.
6. Units of the measurement should be clearly stated along the titles or headings.
7. Abbreviations and symbols should be avoided as far as possible.
8. Sources of the data should be given at the bottom of the data.
9. In case irregularities creep in table or any feature is not sufficiently explained, references and foot notes must be given.
10. The rounding of figures should be unbiased.

Classification

"Classified and arranged facts speak of themselves, and narrated they are as dead as mutton" This quote is given by J.R. Hicks.

The process of dividing the data into different groups (viz. classes) which are homogeneous within but heterogeneous between themselves, is called a classification.

It helps in understanding the salient features of the data and also the comparison with similar data. For a final analysis it is the best friend of a statistician.

Methods of Classification

The data is classified in the following ways :

1. According to attributes or qualities this is divided into two parts :

   (A) Simple classification

   (B) Multiple classification.

2. According to variable or quantity or classification according to class intervals. -

Qualitative Classification : When facts are grouped according to the qualities (attributes) like religion, literacy, business etc., the classification is called as qualitative classification.

(A) Simple Classification: It is also known as classification according to Dichotomy. When data (facts) are divided into groups according to their qualities, the classification is called as 'Simple Classification'. Qualities are denoted by capital letters (A, B, C, D ......) while the absence of these qualities are denoted by lower case letters (a, b, c, d, ... etc.) For example ,

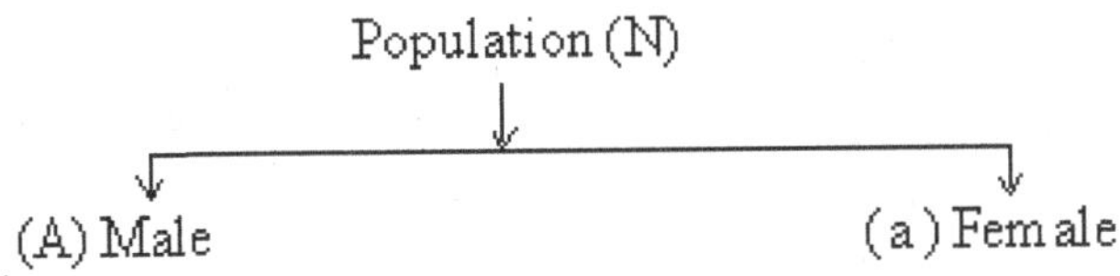

(B) Manifold or multiple classification : In this method data is classified using one or more qualities. First, the data is divided into two groups (classes) using one of the qualities. Then using the remaining qualities, the data is divided into different subgroups. For example, the population of a country is classified using three attributes: sex, literacy and business as,

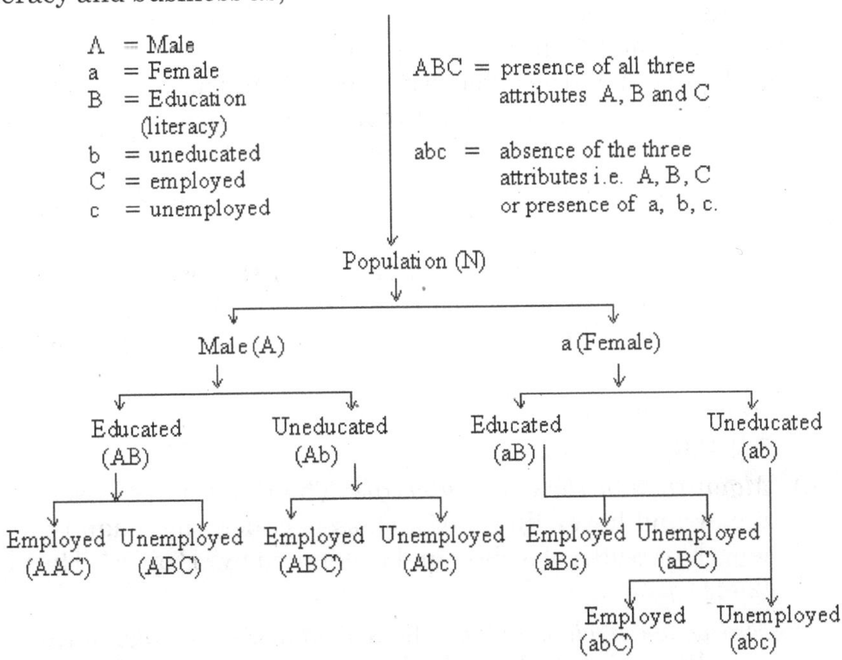

Classification according to class intervals or variables : The data which is expressed in numbers (quantitative data), is classified according to class-intervals. While forming class-intervals one should bear in mind that each and every item must be covered. After finding the least value of an item and the highest value of an item, classify these items into different class-intervals. For example if in any data the age of 100 persons ranging from 2 years to 47 years, is given, then the classification of this data can be done in this way:.

Table - 1

Age in years	No. of persons
0 - 10	5
10 - 20	9
20 - 30	32
30 - 40	34
40 - 50	20

In deciding on the grouping of the data into classes, for the purpose of reducing it to a manageable form, we observe that the number of classes should not be too large. If it were so then the object of summarization would be defeated. The number of classes should also not be too small because then we will miss a great deal of detail available and get a distorted picture. As a rule one should have between 10 and 25 classes, the actual number depending on the total frequency. Further, classes should be exhaustive; they should not be overlapping, so that no observed value falls in more than one class. Apart from exceptions, all classes should have the same length.

According to the class-intervals in classification the following terms are used:

i) *Class-limits:* A class is formed within the two values. These values are known as the class-limits of that class. The lower value is called the lower limit and is denoted by l1 while the higher value is called the upper limit of the class and is denoted by l2. In the example given above, the first class-interval has l1 = 0 and l2 = 10.

ii) *Magnitude of the class-intervals:* The difference between the upper and lower limits of a class is called the magnitude or length or width of a class and is denoted by ' i ' or ' c '. Thus i ° (l2 - l1).

iii) *Mid-value or class-mark:* The arithmetical average of the two class limits (i.e. the lower limit and the upper limit) is called the mid-value or the class mark of that class-interval. For example, the mid-value of the class-interval (0 - 10) is and so on.

iv) *Class frequency:* The units of the data belong to any one of the groups or classes. The total number of these units is known as the frequency of that class and is denoted by fi or simply f. In the above example, the frequencies of the classes in the given order are 5, 9, 32, 34 and 40 respectively.

Classification is of two types according to the class-intervals - (i) Exclusive Method (ii) Inclusive Method.

i) *Exclusive Method:* In this method the upper limit of a class becomes the lower limit of the next class. It is called ' Exclusive ' as we do not put any item that is equal to the upper limit of a class in the same class; we put it in the next class, i.e. the upper limits of classes are excluded from them. For example, a person of age 20 years will not be included in the class-interval (10 - 20) but taken in the next class (20 - 30), since in the class interval (10 - 20) only units ranging from 10 - 19 are included. The exclusive-types of class-intervals can also be expressed as :0 and below 10 or 0 - 9.9 10 and below 20 or 10 - 19.9 20 and below 30 or 20 - 29.9 and so on.

ii) *Inclusive Method:* In this method the upper limit of any class interval is kept in the same class-interval. In this method the upper limit of a previous class is less by 1 from the lower limit of the next class interval. In short this method allows a class-interval to include both its lower and upper limits within it. For example:

Table - 2

Inclusive method		Inclusive method	
Class	Frequency	Class	Frequency
0 - 4	5	0 - 4.9	5
5 - 9	7	5 - 9.9	7
10 - 14	9	10 - 14.9	9
15 - 19	12	15 - 19.9	12
20 - 24	11	20 - 24.9	11
25 - 29	14	25 - 29.9	14

Class Boundaries: Weights are recorded to the nearest Kg The class-intervals 60 - 62 includes all measurements from 59.50000... to 62.50000 ... Kg ; the variable being a continuous one. These numbers, indicated briefly by the exact numbers 59.5 and 62.5, are called class-boundaries or true class limits. The smaller number 59.5 is the lower class boundary and the larger one 62.5 is the upper class boundary.

In any problem if the class-intervals are given as the inclusive type, then they should first be converted into the exclusive-type . For this we require a correction factor.

Correction factor = $\frac{1}{2}$ (the upper limit of a class - the lower limit of the next class) which is generally 0.5.

Now you subtract it from the lower limits and add it to the upper limits of the class-intervals given in the inclusive-method. The class-intervals given above can be written after correction as :

Class	Frequency
0 - 4.5	5
4.5 - 9.5	7
9.5 - 14.5	9
14.5 - 19.5	12
19.5 - 24.5	11
24.5 - 29.5	14

To obtain class-intervals when their mid-values are given, use the following formulae:

Lower limit (l_1) = m - $i/_2$ and upper limit (l_2) = m + $i/_2$

where m = mid-value and i = class-length.

For example, we are given some mid-values as 72, 77, 82, 87, ... Now, consider the first mid-value 72 and also the differences between successive mid-values.

We have 77 - 72 = 5, 82 - 77 = 5, 87 - 82 = 5 ....

which gives the class-length i = 5.

For the first class-interval, l1 = m - $i/_2$ = 72 - $5/_2$ = 69.5 and l2 = 72 + $5/_2$ = 74.5.

Thus the first class-interval is 69.5 - 74.5

and other class-intervals then are 74.5 - 79.5, 79.5 - 84.5, 84.5 - 89.5 ....

Open-end Class Intervals : In any question when the lower limit of the first class-interval or the upper limit of the last class-interval, are not given then subtract the class length of the next immediate class-interval from the upper limit. This will give us the lower limit of the first class-interval. Similarly add the same class length to the lower limit of the last class-interval. But always notice that the lower limit of the first class (i.e. the lowest class) must not be negative or less than 0. For example :

Table - 3

With open ends	Completed classes	With open ends	Completed classes
Below 10	0 - 10	Below 10	0 - 10
10 - 20	10 - 20	10 - 25	10 - 25
20 - 30	20 - 30	25 - 40	25 - 40
30 - 40	30 - 40	40 - 70	40 - 70
40 - 50	40 - 50	above 70	70 - 100
above 50	50 - 60		

Relative Frequency Distribution

The relative frequency of a class is the frequency of the class divided by the total number of frequencies of the class and is generally expresses as a percentage.

Example The weight of 100 persons were given as under :

Weights (in Kgs)	60 - 62	63 - 63	66 - 68	69 - 71	72 - 74
No. of persons (f_i)	5	18	42	27	8

Solution:

Table - 4

Relative frequency table		
Weight (in Kg.)	No. of persons (f_i)	Relative frequency
60 - 62	5	$\frac{5}{100} \times 5\%$ or 0.05
63 - 65	18	$\frac{18}{100} \times 18\%$ or 0.18
66 - 68	42	$\frac{42}{100} \times 42\%$ or 0.42
69 - 71	27	$\frac{27}{100} \times 27\%$ or 0.27
72 - 74	8	$\frac{8}{100} \times 8\%$ or 0.08
Total	$\Sigma f_i = 100$	

Note: The word frequency of a class means, the number of times the class is repeated in the data or the total number of items or observations of the data belongs to that class.

Cumulative Frequency

Many a times the frequencies of different classes are not given. Only their cumulative frequencies are given. The total frequency of all values less than or equal to the upper class boundary of a given class-interval is called the cumulative frequency up to and including that class interval. In this situation both the limits of a class-interval are not written; either lower or upper limit is written. These cumulative frequencies are called less than or more than cumulative frequencies. For example ,

Class-interval	0 - 10	10 - 20	20 - 30	30 - 40	40 - 50
Frequency	4	9	5	12	15

Table - 5

Less than cumulative frequency		More than cumulative frequency	
(Upper limits)	(cum. freq.)	(Lower limits)	(cum. freq.)
Less than 10	4	More than 0	45 = 41 + 4
Less than 20	4 +9 = 13	More than 10	41 = 32 + 9
Less than 30	13 + 5 = 18	More than 20	32 = 27 + 5
Less than 40	18 + 12= 30	More than 30	27 = 15 +12
Less than 50	30 + 15 = 45	More than 40	15

Preparation of Frequency Distribution

We shall now study how to classify the raw data in a tabular form. Consider the data collected by one of the surveyors, interviewing about 50 people. This is as follows :

Size of the Shoes : 2, 5, 6, 8, 2, 5, 6, 7, 6, 8, 7, 4, 3, .. This is called the raw data. Here some values repeat themselves. For instance the size 5 is repeated 10 times in 50 people. We say that the value of 5 of the variate has the frequency of 10. Frequency means the number of times a value of the variate or an attribute, as the case may be, is repeated in the data. A table which shows each value of the characteristic with its corresponding frequency, is known as a Frequency Distribution. The procedure of preparing such a table is explained as below :

Discrete variate : Consider the raw data which gives the size of shoes of 30 persons

2, 5, 6, 4, 5, 7, 4, 4, 6, 2

3, 5, 5, 4, 5, 6, 5, 4, 3, 2

4, 4, 5, 4, 5, 5, 3, 2, 4, 4

The least value is 2 and the highest is 7. All sizes are integers between 2 and 7 (both inclusive). We can prepare a frequency distribution table as follows :

Table - 6

Sizes of shoes	Tally Marks	Frequency
2	||||	4
3	|||	3
4	~~||||~~ ~~||||~~	10
5	~~||||~~ ~~||||~~	9
6	|||	3
7	|	1
Total		30

In this example the size difference from 2 to 7 is very small. If the range of a variate is very large, it is inconvenient to prepare a frequency distribution for each value of the variate. In such a case we divide the variate into convenient groups and prepare a table showing the groups and their corresponding frequencies. Such a table is called a grouped frequency distribution.

Consider the marks (out of 100) of 50 students as below :

40, 39, 43, 62, 30, 47, 33, 31, 17, 28

36, 29, 40, 32, 39, 24, 57, 42, 15, 30

50, 52, 47, 65, 31, 07, 37, 47, 17, 20

25, 53, 65, 85, 89, 56, 55, 41, 43, 10

44, 40, 69, 22, 40, 65, 39, 36, 71, 12

The range of the variate (marks) is very large. Also we are eager to know the performance of the students. The passing limit is 35 and above. Marks between 35 and 44 form the third class (or grade). Marks ranging between 45 - 59 are considered as second class and 60 - 100 form the first class. Thus we have a grouped frequency distribution as:

Table - 7

Marks	Tally Marks	Frequency
0 - 34	𝍸 𝍸 𝍸 I	16
35 - 44	𝍸 𝍸 𝍸 III	18
45 - 59	𝍸 IIII	9
60 - 100	𝍸 II	7
Total		50

Diagrammatic and Graphic Displays

In the last chapter we have seen how to condense the mass of data by the method of classification and tabulation. It is not always easy for a layman to understand figures, nor is it is interesting for him. Apart from that too many figures are often confusing. One of the most convincing and appealing ways in which statistical results may be represented is through graphs and diagrams. It is for this reason that diagrams are often used by businessmen, newspapers, magazines, journals, government agencies and also for advertising and educating people.

Bar Diagrams

1) Simple 'Bar diagram':- It represents only one variable. For example sales, production, population figures etc. for various years may be shown by simple bar charts. Since these are of the same width and vary only in heights (or lengths), it becomes very easy for readers to study the relationship. Simple bar diagrams are very popular in practice. A bar chart can be either vertical or horizontal; vertical bars are more popular.

Illustration: The following table gives the birth rate per thousand of different countries over a certain period of time.

Country	Birth rate	Country	Birth rate
India	33	ChinaNew	40
Germany	15	Zealand	30
U. K.	20	Sweden	15

Represent the above data by a suitable diagram.

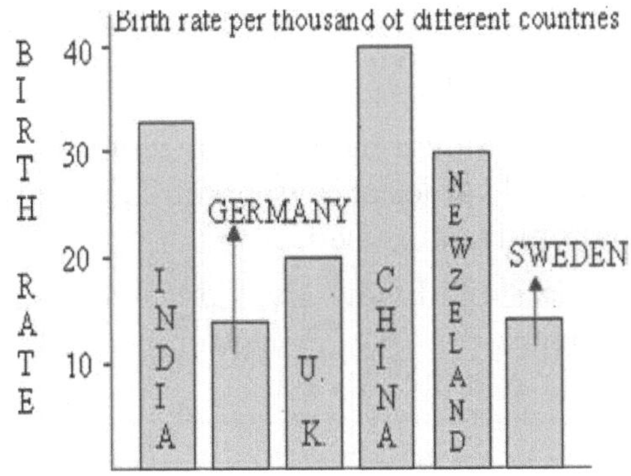

Comparing the size of bars, you can easily see that China's birth rate is the highest while Germany and Sweden equal in the lowest positions. Such diagrams are also known as component bar diagrams.

2) Sub - divided Bar Diagram:- While constructing such a diagram, the various components in each bar should be kept in the same order. A common and helpful arrangement is that of presenting each bar in the order of magnitude with the largest component at the bottom and the smallest at the top. The components are shown with different shades or colours with a proper index.

Illustration: During 1968 - 71, the number of students in University 'X' are as follows. Represent the data by a similar diagram.

Year	*Arts*	*Science*	*Law*	*Total*
1968 - 69	20,000	10,000	5,000	35,000
1969 - 70	26,000	9,000	7,000	42,000
1970 - 71	31,000	9,500	7,500	48,000

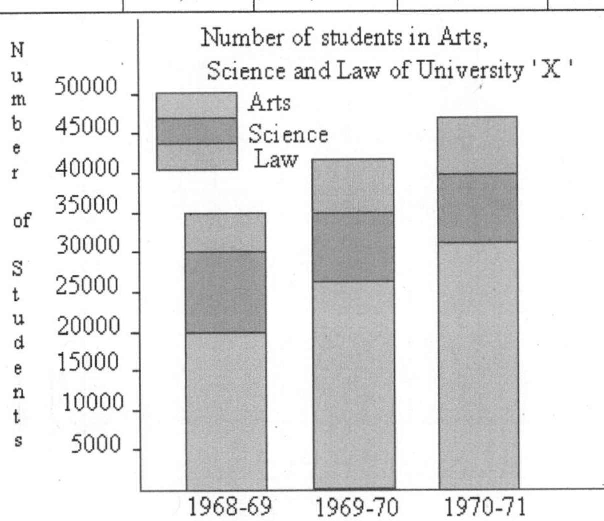

3) Multiple Bar Diagram:- This method can be used for data which is made up of two or more components. In this method the components are shown as separate adjoining bars. The height of each bar represents the actual value of the component. The components are shown by different shades or colours. Where changes in actual values of component figures only are required, multiple bar charts are used.

Illustration:- The table below gives data relating to the exports and imports of a certain country X (in thousands of dollars) during the four years ending in 1930 - 31.

Year	***Export***	***Import***
1927 - 28	319	250
1928 - 29	339	263
1929 - 30	345	258
1930 - 31	308	206

Represent the data by a suitable diagram

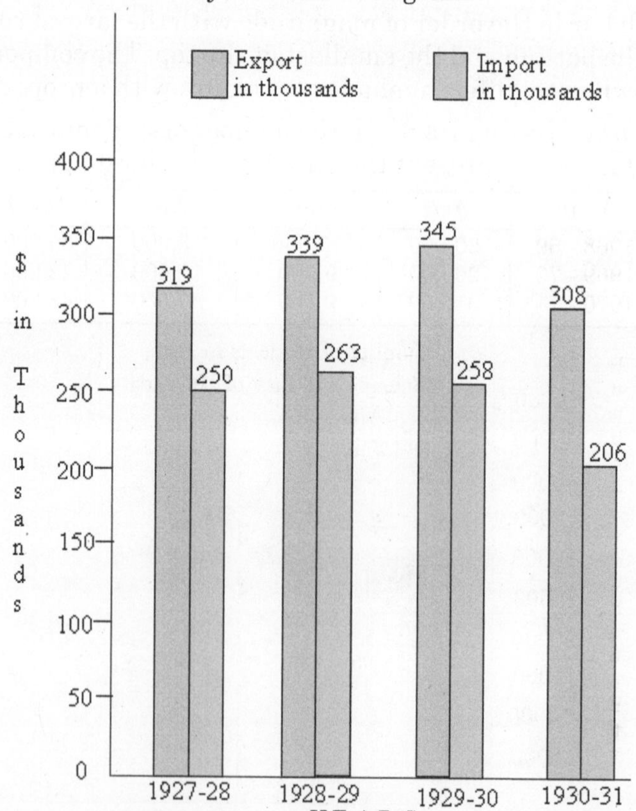

4) Deviation Bar Charts:- Deviation bars are used to represent net quantities - excess or deficit i.e. net profit, net loss, net exports or imports, swings in voting etc. Such bars have both positive and negative values. Positive values lie above the base line and negative values lie below it.

Illustration:-

Years	***Sales***	***Net profits***
1985 - 86	10%	50%
1986 - 87	14%	-20
1987 - 88	12%	-10%

Present the above data by a suitable diagram showing the sales and net profits of private industrial companies.

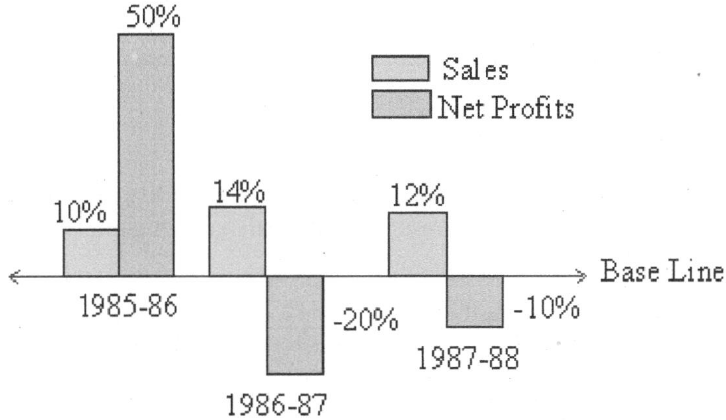

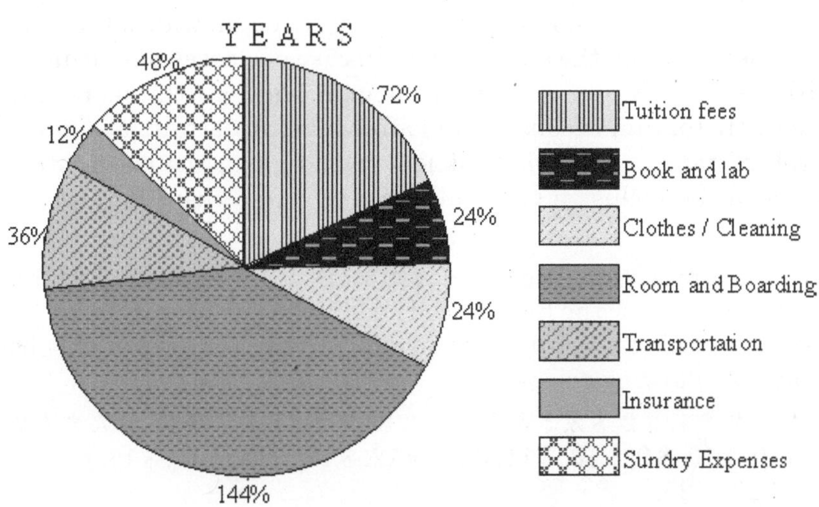

Uses:- A pie diagram is useful when we want to show relative positions (proportions) of the figures which make the total. It is also useful when the components are many in number.

Note:- The sectors of the circle (i.e. of a pie diagram) are ordered from largest to the smallest for easier interpretation of the data and they must be drawn in the counter-clockwise direction.

Graphs

A graph is a visual representation of data by a continuous curve on a squared (graph) paper. Like diagrams, graphs are also attractive, and eye-catching, giving a bird's eye-view of data and revealing their inner pattern.

Graphs of Frequency Distributions:-

The methods used to represent a grouped data are :-

1. Histogram
2. Frequency Polygon
3. Frequency Curve
4. Ogive or Cumulative Frequency Curve

Histogram :- It is defined as a pictorial representation of a grouped frequency distribution by means of adjacent rectangles, whose areas are proportional to the frequencies.

To construct a Histogram, the class intervals are plotted along the x-axis and corresponding frequencies are plotted along the y - axis. The rectangles are constructed such that the height of each rectangle is proportional to the frequency of the that class and width is equal to the length of the class. If all the classes have equal width, then all the rectangles stand on the equal width. In case of classes having unequal widths, rectangles too stand on unequal widths (bases). For open-classes, Histogram is constructed after making certain assumptions. As the rectangles are adjacent leaving no gaps, the class-intervals become of the inclusive type, adjustment is necessary for end points only.

For example, in a book sale, you want to determine which books were most popular, the high priced books, the low priced books, books most neglected etc. Let us say you sold a total 31 books at this book-fair at the following prices.

$ ....2, $ 1, $ 2, $ 2, $ 3, $ 5, $ 6, $ 17, $ 17, $ 7, $ 15, $ 7, $ 7, $ 18, $ 8, $ 10, $ 10, $ 9, $ 13, $ 11, $ 12, $ 12, $ 12, $ 14, $ 16, $ 18, $ 20, $ 24, $ 21, $ 22, $ 25.

The books are ranging from \$1 to \$25. Divide this range into number of groups, class intervals. Typically, there should not be fewer than 5 and more than 20 class-intervals are best for a frequency Histogram.

Our first class-interval includes the lowest price of the data and, the last-interval of course includes, the highest price. Also make sure that overlapping is avoided, so that, no one price falls into two class-intervals. For example you have class intervals as 0-5, 5-10, 10-15 and so on, then the price \$10 falls in both 5-10 and 10-15. Instead if we use \$1 - \$5, \$6=\$10, the class-intervals will be mutually exclusive.

Therefore now we have distribution of books at a book-fair

Class-interval	***Frequency***
\$ 1- \$ 5	6
\$6 - \$10	8
\$11 - \$15	10
\$16 - \$20	3
\$21 - \$25	4
Total	n = ? fi = 31

Note that each class-interval is of equal width i.e. \$5 inclusive. Now we draw the frequency Histogram as under.

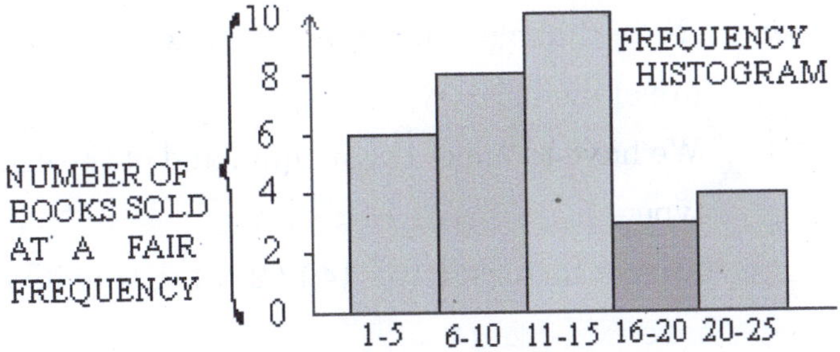

Relative Frequency Histogram:- It uses the same data. The only difference is that it compares each class-interval with the total number of items i.e. instead of the frequency of each class-interval, their relative frequencies are used. Naturally the vertical axis (i.e. y-axis) uses the relative frequencies in places of frequencies.

In the above case we have

Class-interval	*Frequency*	*Relative frequency*
$ 1 - $ 5	6	6/31
$ 6 - $10	8	8/31
$11 - $15	10	10/31
$16 - $20	3	3/31
$21 - $25	4	4/31

The Histogram is same as in above case.

Construction of Histogram when class-intervals are unequal:- In a Histogram, a rectangle is proportional to the frequency of the concern class-interval. Naturally, if the class-intervals are of unequal widths, we have to adjust the heights of the rectangle accordingly. We know that the area of a rectangle = l. h. Now suppose the width (l) of a class is double that of a normal class interval, its height and thus the corresponding frequency must be halved. After this precaution has been taken, the construction of the Histogram of classes of unequal intervals is the same as before.

Note :- The smallest class-interval should be assumed to be " NORMAL "

Illustration:- Represent the following data by means of Histogram.

Classes :	11-14	16-19	21-24	26-29	31-39	41-59	61-79
Frequencies :	7	19	27	15	12	12	8

Solution: Note that class-intervals are unequal and also they are of inclusive type.

We have to make them equal and of the exclusive type.

Correct factor = (16 - 14) / 2 = 1. Using it we have

Classes :	10-15	15-20	20-25	25-30	30-40	40-60	60-80
Frequencies :	7	19	27	15	12	12	8
Adjusted Heights : (Frequencies)	7	19	27	15	12/2 = 6	12/4 = 3	12/4 = 3

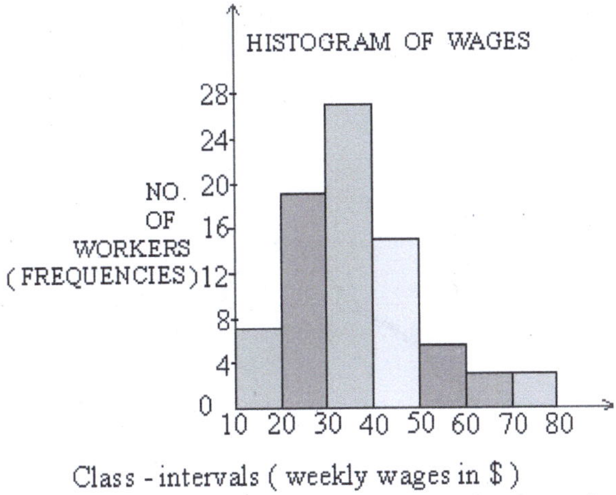

2) Frequency Polygon:- Here the frequencies are plotted against the mid-points of the class-intervals and the points thus obtained are joined by line segments.

Example

Height in cm.	150 - 154	154 - 158	158 - 162	162 - 166	166 - 170
No. of children	10	15	20	12	8

The polygon is closed at the base by extending it on both its sides (ends) to the midpoints of two hypothetical classes, at the extremes of the distribution, with zero frequencies.

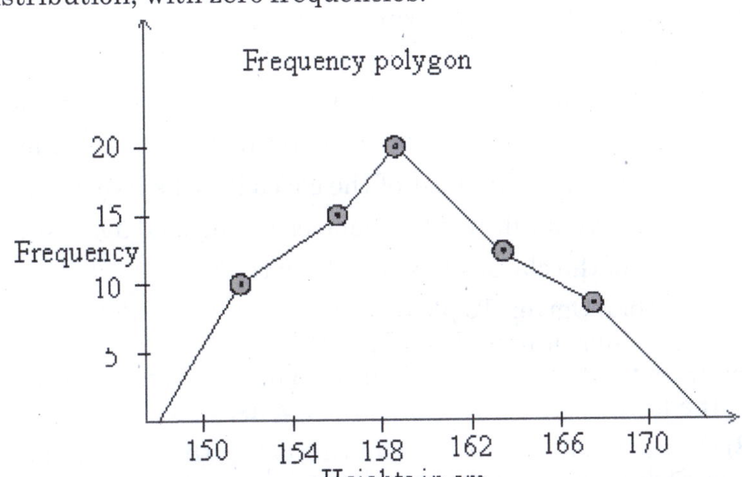

On comparing the Histogram and a frequency polygon, you will notice that, in frequency polygons the points replace the bars (rectangles). Also, when several distributions are to be compared on the same graph paper, frequency polygons are better than Histograms.

3) Frequency Distribution (Curve):- Frequency distribution curves are like frequency polygons. In frequency distribution, instead of using straight line segments, a smooth curve is used to connect the points. The frequency curve for the above data is shown as:

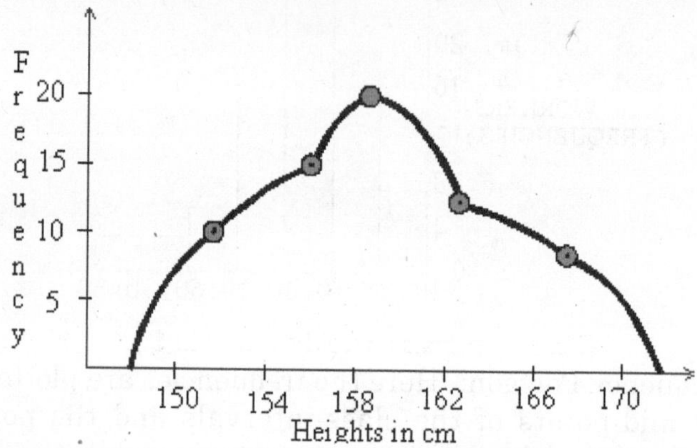

4) Ogives or Cumulative Frequency Curves:- When frequencies are added, they are called cumulative frequencies. The curve obtained by plotting cumulating frequencies is called a cumulative frequency curve or an ogive (pronounced ojive).

To construct an Ogive:-

1) Add up the progressive totals of frequencies, class by class, to get the cumulative frequencies.
2) Plot classes on the horizontal (x-axis) and cumulative frequencies on the vertical (y-axis).
3) Join the points by a smooth curve. Note that Ogives start at (i) zero on the vertical axis, and (ii) outside class limit of the last class. In most of the cases it looks like 'S'.

   Note that cumulative frequencies are plotted against the 'limits' of the classes to which they refer.

(A) Less than Ogive:- To plot a less than ogive, the data is arranged in ascending order of magnitude and the frequencies are cumulated starting from the top. It starts from zero on the y-axis and the lower limit of the lowest class interval on the x-axis.

(B) Greater than Ogive:- To plot this ogive, the data are arranged in the ascending order of magnitude and frequencies are cumulated from the bottom. This curve ends at zero on the y-axis and the upper limit of the highest class interval on the x-axis.

Illustrations:- On a graph paper, draw the two ogives for the data given below of the I.Q. of 160 students.

Class -intervals :	60 - 70	70 - 80	80 - 90	90 - 100	100 - 110
No. of students :	2	7	12	28	42
	110 - 120	120 - 130	130 - 140	140 - 150	150 - 160
	36	18	10	4	1

Solution classes	f	c.f. less than	c.f. greater tha
60 - 70	2	2	159+1 = 160
70 - 80	7	2+7 = 9	151+7 = 158
80 - 90	12	9+12 = 21	139+12 = 151
90 - 100	28	21+28 = 49	111+28 = 139
100 - 110	42	49+4 = 91	69+42 = 111
110 - 120	36	91+36 = 127	33+36 = 69
120 - 130	18	127+18 = 145	15+18 = 33
130 - 140	10	145+10 = 155	5+10 = 15
140 - 150	4	155+4 = 159	1+4 = 5
150 - 160	1	159+1 = 160	1
	$\Sigma f = 160$		

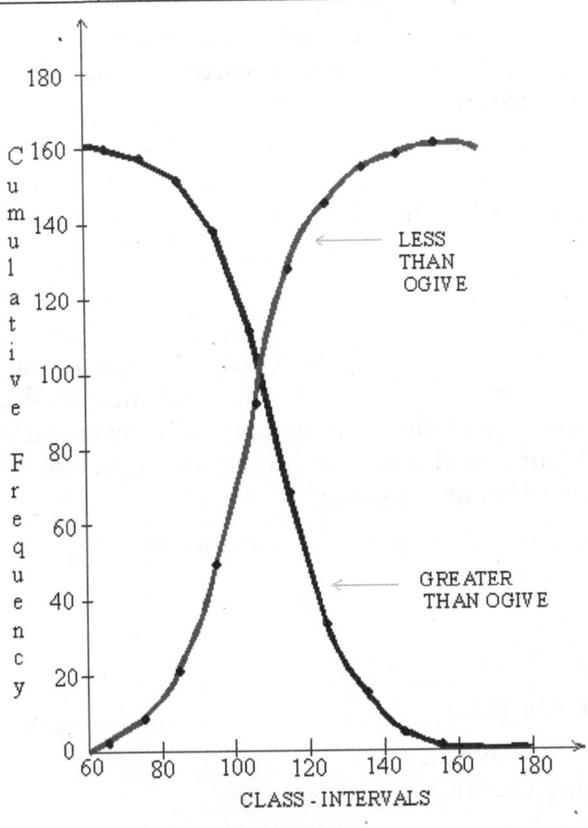

Uses: Certain values like median, quartiles, deciles, quartile deviation, coefficient of skewness etc. can be located using ogives. It can be used to find the percentage of items having values less than or greater than certain value. Ogives are helpful in the comparison of the two distributions.

Measures of Central Tendency

Introduction

In the previous chapter, we have studied how to collect raw data, its classification and tabulation in a useful form, which contributes in solving many problems of statistical concern. Yet, this is not sufficient, for in practical purposes, there is need for further condensation, particularly when we want to compare two or more different distributions. We may reduce the entire distribution to one number which represents the distribution.

A single value which can be considered as typical or representative of a set of observations and around which the observations can be considered as Centered is called an 'Average' (or average value) or a Center of location. Since such typical values tends to lie centrally within a set of observations when arranged according to magnitudes, averages are called measures of central tendency.

In fact the distribution have a typical value (average) about which, the observations are more or less symmetrically distributed. This is of great importance, both theoretically and practically. Dr. A.L. Bowley correctly stated, "Statistics may rightly be called the science of averages."

The word average is commonly used in day-to-day conversations. For example, we may say that Abert is an average boy of my class; we may talk of an average American, average income, etc. When it is said, "Abert is an average student," it means is that he is neither very good nor very bad, but a mediocre student. However, in statistics the term average has a different meaning.

The fundamental measures of tendencies are:

(1) Arithmetic mean

(2) Median

(3) Mode

(4) Geometric mean

(5) Harmonic mean

(6) Weighted averages

However the most common measures of central tendencies or Locations are Arithmetic mean, median and mode. We therefore, consider the Arithmetic mean

Arithmetic Mean

This is the most commonly used average which you have also studied and used in lower grades. Here are two definitions given by two great masters of statistics.

Horace Sacrist : Arithmetic mean is the amount secured by dividing the sum of values of the items in a series by their number.

W.I. King : The arithmetic average may be defined as the sum of aggregate of a series of items divided by their number.

Thus, the students should add all observations (values of all items) together and divide this sum by the number of observations (or items).

Chapter 7

Regression Analysis

In statistics, regression analysis is a statistical process for estimating the relationships among variables. It includes many techniques for modelling and analyzing several variables, when the focus is on the relationship between a dependent variable and one or more independent variables. More specifically, regression analysis helps one understand how the typical value of the dependent variable (or 'criterion variable') changes when any one of the independent variables is varied, while the other independent variables are held fixed. Most commonly, regression analysis estimates the conditional expectation of the dependent variable given the independent variables – that is, the average value of the dependent variable when the independent variables are fixed. Less commonly, the focus is on a quantile, or other location parameter of the conditional distribution of the dependent variable given the independent variables. In all cases, the estimation target is a function of the independent variables called the regression function. In regression analysis, it is also of interest to characterize the variation of the dependent variable around the regression function which can be described by a probability distribution.

Regression analysis is widely used for prediction and forecasting, where its use has substantial overlap with the field of machine learning. Regression analysis is also used to understand which among the independent variables are related to the dependent variable, and to explore the forms of these relationships. In restricted circumstances, regression analysis can be used to infer causal relationships between the independent and dependent variables. However this can lead to illusions or false relationships, so caution is advisable; for example, correlation does not imply causation.

Many techniques for carrying out regression analysis have been developed. Familiar methods such as linear regression and ordinary least squares regression are parametric, in that the regression function is defined in terms of a finite number of unknown parameters that are estimated from the data. Nonparametric regression refers to techniques that allow the regression function to lie in a specified set of functions, which may be infinite-dimensional.

The performance of regression analysis methods in practice depends on the form of the data generating process, and how it relates to the regression approach being used. Since the true form of the data-generating process is generally not known, regression analysis often depends to some extent on making assumptions about this process. These assumptions are sometimes testable if a sufficient quantity of data is available. Regression models for prediction are often useful even when the assumptions are moderately violated, although they may not perform optimally. However, in many applications, especially with small effects or questions of causality based on observational data, regression methods can give misleading results.

Regression Models

Regression models involve the following variables:

- The unknown parameters, denoted as β, which may represent a scalar or a vector.
- The independent variables, X.
- The dependent variable, *Y*.

In various fields of application, different terminologies are used in place of dependent and independent variables.

A regression model relates *Y* to a function of X and β.

$$Y \approx f(\mathbf{X}, \boldsymbol{\beta})$$

The approximation is usually formalized as $E(Y \mid X) = f(X, \beta)$. To carry out regression analysis, the form of the function f must be specified. Sometimes the form of this function is based on knowledge about the relationship between *Y* and X that does not rely on the data. If no such knowledge is available, a flexible or convenient form for f is chosen.

Assume now that the vector of unknown parameters β is of length k. In order to perform a regression analysis the user must provide information about the dependent variable *Y*:

- If N data points of the form (Y, X) are observed, where $N < k$, most classical approaches to regression analysis cannot be performed: since the system of equations defining the regression model is underdetermined, there are not enough data to recover β.

- If exactly $N = k$ data points are observed, and the function f is linear, the equations $Y = f(X, \beta)$ can be solved exactly rather than approximately. This reduces to solving a set of N equations with N unknowns (the elements of β), which has a unique solution as long as the X are linearly independent. If f is nonlinear, a solution may not exist, or many solutions may exist.
- The most common situation is where $N > k$ data points are observed. In this case, there is enough information in the data to estimate a unique value for β that best fits the data in some sense, and the regression model when applied to the data can be viewed as an overdetermined system in β.

In the last case, the regression analysis provides the tools for:

1. Finding a solution for unknown parameters β that will, for example, minimize the distance between the measured and predicted values of the dependent variable Y (also known as method of least squares).
2. Under certain statistical assumptions, the regression analysis uses the surplus of information to provide statistical information about the unknown parameters β and predicted values of the dependent variable Y.

Necessary Number of Independent Measurements

Consider a regression model which has three unknown parameters, β_0, β_1, and β_2. Suppose an experimenter performs 10 measurements all at exactly the same value of independent variable vector X (which contains the independent variables X_1, X_2, and X_3). In this case, regression analysis fails to give a unique set of estimated values for the three unknown parameters; the experimenter did not provide enough information. The best one can do is to estimate the average value and the standard deviation of the dependent variable Y. Similarly, measuring at two different values of X would give enough data for a regression with two unknowns, but not for three or more unknowns.

If the experimenter had performed measurements at three different values of the independent variable vector X, then regression analysis would provide a unique set of estimates for the three unknown parameters in β.

In the case of general linear regression, the above statement is equivalent to the requirement that the matrix X^TX is invertible.

Satistical Assumptions

When the number of measurements, N, is larger than the number of unknown parameters, k, and the measurement errors ε_i are normally distributed then *the excess of information* contained in $(N - k)$ measurements is used to make statistical predictions about the unknown parameters. This excess of information is referred to as the degrees of freedom of the regression.

Underlying Assumptions

Classical assumptions for regression analysis include:

- The sample is representative of the population for the inference prediction.
- The error is a random variable with a mean of zero conditional on the explanatory variables.
- The independent variables are measured with no error. (Note: If this is not so, modelling may be done instead using errors-in-variables model techniques).
- The predictors are linearly independent, i.e. it is not possible to express any predictor as a linear combination of the others.
- The errors are uncorrelated, that is, the variance–covariance matrix of the errors is diagonal and each non-zero element is the variance of the error.

The variance of the error is constant across observations (homoscedasticity). If not, weighted least squares or other methods might instead be used.

These are sufficient conditions for the least-squares estimator to possess desirable properties; in particular, these assumptions imply that the parameter estimates will be unbiased, consistent, and efficient in the class of linear unbiased estimators. It is important to note that actual data rarely satisfies the assumptions. That is, the method is used even though the assumptions are not true. Variation from the assumptions can sometimes be used as a measure of how far the model is from being useful. Many of these assumptions may be relaxed in more advanced treatments. Reports of statistical analyses usually include analyses of tests on the sample data and methodology for the fit and usefulness of the model.

Assumptions include the geometrical support of the variables. Independent and dependent variables often refer to values measured at point locations. There may be spatial trends and spatial autocorrelation in the variables that violate statistical assumptions of regression.

Geographic weighted regression is one technique to deal with such data. Also, variables may include values aggregated by areas. With aggregated data the modifiable areal unit problem can cause extreme variation in regression parameters. When analyzing data aggregated by political boundaries, postal codes or census areas results may be very distinct with a different choice of units.

Linear Regression

In linear regression, the model specification is that the dependent variable, y_i is a linear combination of the *parameters* (but need not be linear in the *independent variables*). For example, in simple linear regression for modelling n data points there is one independent variable: x_i, and two parameters, β_0 and β_1:

$$\text{straight line: } y_i = \beta_0 + \beta_1 x_i + \varepsilon_i, \quad i = 1, \ldots, n.$$

In multiple linear regression, there are several independent variables or functions of independent variables.

Adding a term in x_i^2 to the preceding regression gives:

$$\text{parabola: } y_i = \beta_0 + \beta_1 x_i + \beta_2 x_i^2 + \varepsilon_i, i = 1, \ldots, n.$$

This is still linear regression; although the expression on the right hand side is quadratic in the independent variable x_i, it is linear in the parameters β_0, and

In both cases, ε_i is an error term and the subscript i indexes a particular observation.

Given a random sample from the population, we estimate the population parameters and obtain the sample linear regression model:

$$\widehat{y_i} = \hat{\beta}_0 + \hat{\beta}_1 x_i.$$

The residual, $e_i = y_i - \hat{y}_i$, is the difference between the value of the dependent variable predicted by the model, $\widehat{y_i}$, and the true value of the dependent variable, y_i. One method of estimation is ordinary least squares. This method obtains parameter estimates that minimize the sum of squared residuals, SSE, also sometimes denoted RSS:

$$SSE = \sum_{i=1}^{n} e_i^2.$$

Minimization of this function results in a set of normal equations, a set of simultaneous linear equations in the parameters, which are solved to yield the parameter estimators, $\hat{\beta}_0, \hat{\beta}_1$.

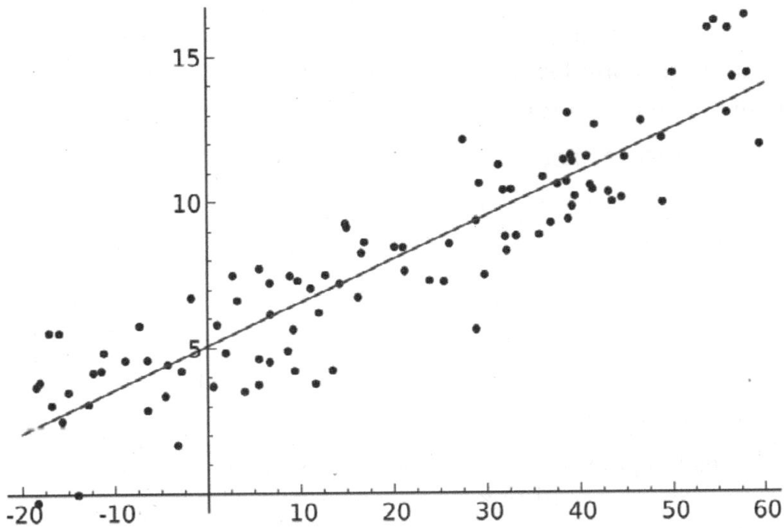

***Figure:** Illustration of linear regression on a data set.*

In the case of simple regression, the formulas for the least squares estimates are

$$\widehat{\beta_1} = \frac{\sum (x_i - \overline{x})(y_i - \overline{y})}{\sum (x_i - \overline{x})^2} \text{ and } \widehat{\beta_0} = \overline{y} - \widehat{\beta_1}\overline{x}$$

where $\overline{x}$ is the mean (average) of the $x$values and y is the mean of the y values.

Under the assumption that the population error term has a constant variance, the estimate of that variance is given by:

$$\hat{\sigma}_\varepsilon^2 = \frac{SSE}{n-2}.$$

This is called the mean square error (MSE) of the regression. The denominator is the sample size reduced by the number of model parameters estimated from the same data, (n-p) for p regressors or (n-p-1) if an intercept is used. In this case, p=1 so the denominator is n-2.

The standard errors of the parameter estimates are given by

$$\hat{\sigma}_{\beta_0} = \hat{\sigma}_\varepsilon \sqrt{\frac{1}{n} + \frac{\overline{x}^2}{\sum (x_i - \overline{x})^2}}$$

$$\hat{\sigma}_{\beta_1} = \hat{\sigma}_{\varepsilon}\sqrt{\frac{1}{\sum(x_i - \bar{x})^2}}.$$

Under the further assumption that the population error term is normally distributed, the researcher can use these estimated standard errors to create confidence intervals and conduct hypothesis tests about the population parameters.

General Linear Model

In the more general multiple regression model, there are p independent variables:

$$y_i = \beta_1 x_{i1} + \beta_2 x_{i2} + \cdots + \beta_p x_{ip} + \varepsilon_i,$$

where x_{ij} is the ith observation on the jth independent variable, and where the first independent variable takes the value 1 for all i (so β_1 is the regression intercept).

The least squares parameter estimates are obtained from p normal equations. The residual can be written as

$$\varepsilon_i = y_i - \hat{\beta}_1 x_{i1} - \cdots - \hat{\beta}_p x_{ip}.$$

The normal equations are

$$\sum_{i=1}^{n}\sum_{k=1}^{p} X_{ij}X_{ik}\hat{\beta}_k = \sum_{i=1}^{n} X_{ij}y_i,\ j = 1,\ldots,p.$$

In matrix notation, the normal equations are written as

$$(\mathbf{X}^{\top}\mathbf{X})\hat{\boldsymbol{\beta}} = \mathbf{X}^{\top}\mathbf{Y},$$

where the ij element of X is x_{ij}, the i element of the column vector Y is y_i, and the j element of $\hat{\beta}$ is $\hat{\beta}_j$. Thus X is $n\times p$, Y is $n\times 1$, and is $p\times 1$. The solution is

Diagnostics

Once a regression model has been constructed, it may be important to confirm the goodness of fit of the model and the statistical significance of the estimated parameters. Commonly used checks of goodness of fit include the R-squared, analyses of the pattern of residuals and hypothesis testing. Statistical significance can be checked by an F-test of the overall fit, followed by t-tests of individual parameters.

Interpretations of these diagnostic tests rest heavily on the model assumptions. Although examination of the residuals can be used to invalidate a model, the results of a t-test or F-test are sometimes more difficult to interpret if the model's assumptions are violated. For example, if the error term does not have a normal distribution, in small samples the estimated parameters will not follow normal distributions and complicate inference. With relatively large samples, however, a central limit theorem can be invoked such that hypothesis testing may proceed using asymptotic approximations.

"Limited Dependent" Variables

The phrase "limited dependent" is used in econometric statistics for categorical and constrained variables.

The response variable may be non-continuous ("limited" to lie on some subset of the real line). For binary (zero or one) variables, if analysis proceeds with least-squares linear regression, the model is called the linear probability model. Nonlinear models for binary dependent variables include the probit and logit model. The multivariate probit model is a standard method of estimating a joint relationship between several binary dependent variables and some independent variables. For categorical variables with more than two values there is the multinomial logit. For ordinal variables with more than two values, there are the ordered logit and ordered probit models. Censored regression models may be used when the dependent variable is only sometimes observed, and Heckman correction type models may be used when the sample is not randomly selected from the population of interest. An alternative to such procedures is linear regression based on polychoric correlation (or polyserial correlations) between the categorical variables. Such procedures differ in the assumptions made about the distribution of the variables in the population. If the variable is positive with low values and represents the repetition of the occurrence of an event, then count models like the Poisson regression or the negative binomial model may be used instead.

Probability Theory

Probability theory is the branch of mathematics concerned with probability, the analysis of random phenomena. The central objects of probability theory are random variables, stochastic processes, and events: mathematical abstractions of non-deterministic events or measured quantities that may either be single occurrences or evolve over time in an apparently random fashion. If an individual coin toss or the roll of dice is considered to be a random event, then if repeated many times the sequence of random events will exhibit certain patterns, which can be studied and predicted. Two representative mathematical results describing such patterns are the law of large numbers and the central limit theorem.

Treatment

Most introductions to probability theory treat discrete probability distributions and continuous probability distributions separately. The more mathematically advanced measure theory based treatment of probability covers both the discrete, the continuous, any mix of these two and more.

Motivation

Consider an experiment that can produce a number of outcomes. The set of all outcomes is called the *sample space* of the experiment. The power set of the sample space is formed by considering all different collections of possible results. For example, rolling an honest die produces one of six possible results. One collection of possible results corresponds to getting an odd number. Thus, the subset {1,3,5} is an element of the power set of the sample space of die rolls. These collections are called *events*. In this case, {1,3,5} is the event that the die falls on some odd number. If the results that actually occur fall in a given event, that event is said to have occurred.

Probability is a way of assigning every "event" a value between zero and one, with the requirement that the event made up of all possible results (in our example, the event {1,2,3,4,5,6}) be assigned a value of one. To qualify as a probability distribution, the assignment of values must satisfy the requirement that if you look at a collection of mutually exclusive events (events that contain no common results, e.g., the events {1,6}, {3}, and {2,4} are all mutually exclusive), the probability that one of the events will occur is given by the sum of the probabilities of the individual events.

The probability that any one of the events {1,6}, {3}, or {2,4} will occur is 5/6. This is the same as saying that the probability of event {1,2,3,4,6} is 5/6. This event encompasses the possibility of any number except five being rolled. The mutually exclusive event {5} has a probability of 1/6, and the event {1,2,3,4,5,6} has a probability of 1, that is, absolute certainty.

Discrete Probability Distributions

Discrete probability theory deals with events that occur in countable sample spaces.

Examples: Throwing dice, experiments with decks of cards, random walk, and tossing coins

Classical definition: Initially the probability of an event to occur was defined as number of cases favourable for the event, over the number of total outcomes possible in an equiprobable sample space.

For example, if the event is "occurrence of an even number when a die is rolled", the probability is given by $\frac{3}{6}=\frac{1}{2}$, since 3 faces out of the 6 have even numbers and each face has the same probability of appearing.

Modern definition: The modern definition starts with a finite or countable set called the sample space, which relates to the set of all *possible outcomes* in classical sense, denoted by Ω. It is then assumed that for each element $x \in \Omega$, an intrinsic "probability" value $f(x)$ is attached, which satisfies the following properties:

1. $f(x) \in [0,1]$ for all $x \in \Omega$;
2. $\sum_{x \in \Omega} f(x) = 1.$

That is, the probability function $f(x)$ lies between zero and one for every value of x in the sample space Ω, and the sum of $f(x)$ over all values x in the sample space Ω is equal to 1. An event is defined as any subset E of the sample space . The probability of the event is defined as

$$P(E) = \sum_{x \in E} f(x).$$

So, the probability of the entire sample space is 1, and the probability of the null event is 0.

The function $f(x)$ mapping a point in the sample space to the "probability" value is called a probability mass function abbreviated as pmf. The modern definition does not try to answer how probability mass functions are obtained; instead it builds a theory that assumes their existence.

Continuous Probability Distributions

Continuous probability theory deals with events that occur in a continuous sample space.

Classical definition: The classical definition breaks down when confronted with the continuous case.

Modern definition: If the outcome space of a random variable X is the set of real numbers ($\mathbb{R}$) or a subset thereof, then a function called the cumulative distribution function (or cdf) F exists, defined by $F(x) = P(X \leq x)$. That is, $F(x)$ returns the probability that X will be less than or equal to x.

The cdf necessarily satisfies the following properties.

1. F is a monotonically non-decreasing, right-continuous function;
2. $\lim_{x \to -\infty} F(x) = 0;$
3. $\lim_{x \to \infty} F(x) = 1.$

If is absolutely continuous, i.e., its derivative exists and integrating the derivative gives us the cdf back again, then the random variable X is said to have a probability density function or pdf or simply density

$$f(x) = \frac{dF(x)}{dx}.$$

For a set $E \subseteq \mathbb{R}$, the probability of the random variable X being in E is

$$P(X \in E) = \int_{x \in E} dF(x).$$

In case the probability density function exists, this can be written as

$$P(X \in E) = \int_{x \in E} f(x) dx.$$

Whereas the *pdf* exists only for continuous random variables, the *cdf* exists for all random variables (including discrete random variables) that take values in

These concepts can be generalized for multidimensional cases on $\mathbb{R}^n$ and other continuous sample spaces.

Measure-Theoretic Probability Theory

The *raison d'être* of the measure-theoretic treatment of probability is that it unifies the discrete and the continuous cases, and makes the difference a question of which measure is used. Furthermore, it covers distributions that are neither discrete nor continuous nor mixtures of the two.

An example of such distributions could be a mix of discrete and continuous distributions—for example, a random variable that is 0 with probability 1/2, and takes a random value from a normal distribution with probability 1/2. It can still be studied to some extent by considering it to have a pdf of $(\delta[x] + \varphi(x)) / 2$, where $\delta[x]$ is the Dirac delta function.

Other distributions may not even be a mix, for example, the Cantor distribution has no positive probability for any single point, neither does it have a density. The modern approach to probability theory solves these problems using measure theory to define the probability space:

Given any set Ω , (also called sample space) and a σ-algebra F on it, a measure P defined on is called a probability measure if

If is the Borel σ-algebra on the set of real numbers, then there is a unique probability measure on for any cdf, and vice versa. The measure corresponding to a cdf is said to be induced by the cdf. This measure coincides with the pmf for discrete variables and pdf for continuous variables, making the measure-theoretic approach free of fallacies.

The $\mathcal{F}$ *probability* of a set in the σ-algebra $\mathcal{F}$ is defined as

$$P(E) = \int_{\omega \in E} \mu_F(d\omega)$$

where the integration is with respect to the measure μ_F induced by

Along with providing better understanding and unification of discrete and continuous probabilities, measure-theoretic treatment also allows us to work on probabilities outside , as in the theory of stochastic processes. For example to study Brownian motion, probability is defined on a space of functions.

Probability Distributions

Certain random variables occur very often in probability theory because they well describe many natural or physical processes. Their distributions therefore have gained *special importance* in probability theory. Some fundamental *discrete distributions* are the discrete uniform, Bernoulli, binomial, negative binomial, Poisson and geometric distributions. Important *continuous distributions* include the continuous uniform, normal, exponential, gamma and beta distributions.

Convergence of Random Variables

In probability theory, there are several notions of convergence for random variables. They are listed below in the order of strength, i.e., any subsequent notion of convergence in the list implies convergence according to all of the preceding notions. Weak convergence: A sequence of random variables $X_1, X_2, \ldots,$ converges weakly to the random variable X if their respective cumulative *distribution functions* $F_1, F_2, \ldots$ converge to the cumulative distribution function of F , wherever is continuous. Weak convergence is also called convergence in distribution.

Most common shorthand notation: $X_n \xrightarrow{\mathcal{D}} X.$

Convergence in probability: The sequence of random $X_1, X_2, \ldots$ variables is said to converge towards the random variable in probability if $\lim_{n\to\infty} P\left(\left|X_n - X\right| \geq \varepsilon\right) = 0$ for every ε > 0.

Most common shorthand notation: $X_n \xrightarrow{P} X$.

Strong convergence: The sequence of random $X_1, X_2, \ldots$ variables is said to converge towards the random variable X strongly if. Strong convergence is also known as almost sure convergence.

Most common shorthand notation:

As the names indicate, weak convergence is weaker than strong convergence. In fact, strong convergence implies convergence in probability, and convergence in probability implies weak convergence. The reverse statements are not always true.

Law of Large Numbers

Common intuition suggests that if a fair coin is tossed many times, then *roughly* half of the time it will turn up *heads*, and the other half it will turn up *tails*. Furthermore, the more often the coin is tossed, the more likely it should be that the ratio of the number of *heads* to the number of *tails* will approach unity. Modern probability provides a formal version of this intuitive idea, known as the law of large numbers. This law is remarkable because it is not assumed in the foundations of probability theory, but instead emerges from these foundations as a theorem. Since it links theoretically derived probabilities to their actual frequency of occurrence in the real world, the law of large numbers is considered as a pillar in the history of statistical theory and has had widespread influence.

The law of large numbers (LLN) states that the sample average

$$\bar{X}_n = \frac{1}{n}\sum_{k=1}^{n} X_k$$

of a sequence of independent and identically distributed random variables X_k converges towards their common expectation μ, provided that the expectation of $|X_k|$ is finite.

It is in the different forms of convergence of random variables that separates the *weak* and the *strong* law of large numbers

Weak law: $\bar{X}_n \xrightarrow{P} \mu$ for $n \to \infty$

Strong law: $\bar{X}_n \xrightarrow{\text{a.s.}} \mu$ for $n \to \infty$.

It follows from the LLN that if an event of probability p is observed repeatedly during independent experiments, the ratio of the observed frequency of that event to the total number of repetitions converges towards p.

For example, if $Y_1, Y_2, \ldots$ are independent Bernoulli random variables taking values 1 with probability *p* and 0 with probability 1-*p*, then $E(Y_i) = p$ for all *i*, so that $\bar{Y}_n$ converges to *p* almost surely.

Central Limit Theorem

"The central limit theorem (CLT) is one of the great results of mathematics." It explains the ubiquitous occurrence of the normal distribution in nature.

The theorem states that the average of many independent and identically distributed random variables with finite variance tends towards a normal distribution *irrespective* of the distribution followed by the original random variables. Formally, let $X_1, X_2, \ldots$ be independent random variables with mean μ and variance $\sigma^2 > 0$. Then the sequence of random variables

$$Z_n = \frac{\sum_{i=1}^{n}(X_i - \mu)}{\sigma\sqrt{n}}$$

converges in distribution to a standard normal random variable.

Notice that for some classes of random variables the classic central limit theorem works rather fast (Berry–Esseen theorem), for example the distributions with finite first, second and third moment from the exponential family, on the other hand for some random variables of the heavy tail and fat tail variety, it works very slow or may not work at all: in such cases one may use the Generalized Central Limit Theorem (GCLT).

Binomial Distribution

In probability theory and statistics, the binomial distribution with parameters *n* and *p* is the discrete probability distribution of the number of successes in a sequence of *n* independent yes/no experiments, each of which yields success with probability *p*. A success/failure experiment is also called a Bernoulli experiment or Bernoulli trial; when *n* = 1, the binomial distribution is a Bernoulli distribution. The binomial distribution is the basis for the popular binomial test of statistical significance.

The binomial distribution is frequently used to model the number of successes in a sample of size *n* drawn with replacement from a population of size *N*. If the sampling is carried out without replacement, the draws are not independent and so the resulting distribution is a hypergeometric distribution, not a binomial one. However, for N much larger than n, the binomial distribution is a good approximation, and widely used.

Specification

Probability Mass Function: In general, if the random variable X follows the binomial distribution with parameters n and p, we write $X \sim B(n, p)$. The probability of getting exactly k successes in n trials is given by the probability mass function:

$$f(k;n,p) = \Pr(X = k) = \binom{n}{k} p^k (1-p)^{n-k}$$

for $k = 0, 1, 2, ..., n$, where

$$\binom{n}{k} = \frac{n!}{k!(n-k)!}$$

is the binomial coefficient, hence the name of the distribution. The formula can be understood as follows: we want k successes (p^k) and $n - k$ failures $(1 - p)^{n-k}$. However, the k successes can occur anywhere among the n trials, and there are $\binom{n}{k}$ different ways of distributing k successes in a sequence of n trials.

In creating reference tables for binomial distribution probability, usually the table is filled in up to $n/2$ values. This is because for $k > n/2$, the probability can be calculated by its complement as

$$f(k,n,p) = f(n-k,n,1-p).$$

Looking at the expression $f(k, n, p)$ as a function of k, there is a k value that maximizes it. This k value can be found by calculating

$$\frac{f(k+1,n,p)}{f(k,n,p)} = \frac{(n-k)p}{(k+1)(1-p)}$$

and comparing it to 1. There is always an integer M that satisfies

$$(n+1)p - 1 \leq M < (n+1)p.$$

$f(k, n, p)$ is monotone increasing for $k < M$ and monotone decreasing for $k > M$, with the exception of the case where $(n + 1)p$ is an integer. In this case, there are two values for which f is maximal: $(n + 1)p$ and $(n + 1)p - 1$. M is the *most probable* (*most likely*) outcome of the Bernoulli trials and is called the mode. Note that the probability of it occurring can be fairly small.

Recurrence relation

$$\left\{p(n-k)\text{Prob}(k)+(k+1)(p-1)\text{Prob}(k+1)=0, \text{Prob}(0)=(1-p)^n\right\}$$

Cumulative Distribution Function

The cumulative distribution function can be expressed as:

$$F(k;n,p)=\Pr(X\le k)=\sum_{i=0}^{k}\binom{n}{i}p^i(1-p)^{n-i}$$

where is the "floor" under k, i.e. the greatest integer less than or equal to k.

It can also be represented in terms of the regularized incomplete beta function, as follows:

$$\begin{aligned}F(k;n,p) &= \Pr(X\le k)\\ &= I_{1-p}(n-k,k+1)\\ &= (n-k)\binom{n}{k}\int_0^{1-p} t^{n-k-1}(1-t)^k\,dt.\end{aligned}$$

Some closed-form bounds for the cumulative distribution function are given below.

Example

Suppose a biased coin comes up heads with probability 0.3 when tossed. What is the probability of achieving 0, 1,..., 6 heads after six tosses?

$$\Pr(0\text{ heads})=f(0)=\Pr(X=0)=\binom{6}{0}0.3^0(1-0.3)^{6-0}\approx 0.1176$$

$$\Pr(1\text{ head })=f(1)=\Pr(X=1)=\binom{6}{1}0.3^1(1-0.3)^{6-1}\approx 0.3025$$

$$\Pr(2\text{ heads})=f(2)=\Pr(X=2)=\binom{6}{2}0.3^2(1-0.3)^{6-2}\approx 0.3241$$

$$\Pr(3\text{ heads})=f(3)=\Pr(X=3)=\binom{6}{3}0.3^3(1-0.3)^{6-3}\approx 0.1852$$

$$\Pr(4\text{ heads})=f(4)=\Pr(X=4)=\binom{6}{4}0.3^4(1-0.3)^{6-4}\approx 0.0595$$

$$\Pr(5\text{ heads})=f(5)=\Pr(X=5)=\binom{6}{5}0.3^5(1-0.3)^{6-5}\approx 0.0102$$

$$\Pr(6 \text{ heads}) = f(6) = \Pr(X = 6) = \binom{6}{6} 0.3^6 (1-0.3)^{6-6} \approx 0.0007$$

Mean and Variance

If $X \sim B(n, p)$, that is, X is a binomially distributed random variable, n being the total number of experiments and p the probability of each experiment yielding a successful result, then the expected value of X is

$$E[X] = np,$$

(For example, if n=100, and p=1/4, then the average number of successful results will be 25) and the variance

$$\text{Var}[X] = np(1-p).$$

Mode and Median

Usually the mode of a binomial $B(n,p)$ distribution is equal to $(n+1)p$, where is the floor function. However when $(n + 1)p$ is an integer and p is neither 0 nor 1, then the distribution has two modes: $(n + 1)p$ and $(n + 1)p - 1$. When p is equal to 0 or 1, the mode will be 0 and n correspondingly. These cases can be summarized as follows:

$$\text{mode} = \begin{cases} (n+1)p & \text{if } (n+1)p \text{ is 0 or a noninteger,} \\ (n+1)p \text{ and } (n+1)p-1 & \text{if } (n+1)p \in \{1, \ldots, n\}, \\ n & \text{if } (n+1)p = n+1. \end{cases}$$

In general, there is no single formula to find the median for a binomial distribution, and it may even be non-unique. However several special results have been established:

- If np is an integer, then the mean, median, and mode coincide and equal np.
- Any median m must lie within the interval $\#np\# \leq m \leq \#np\#$.
- A median m cannot lie too far away from the mean: $|m - np| \leq \min\{\ln 2, \max\{p, 1-p\}\}$.
- The median is unique and equal to $m = \text{round}(np)$ in cases when either $p \leq 1 - \ln 2$ or $p \leq \ln 2$ or $|m - np| \leq \min\{p, 1-p\}$ (except for the case when $p = ½$ and n is odd).
- When $p = 1/2$ and n is odd, any number m in the interval $½(n - 1) \leq m \leq ½(n + 1)$ is a median of the binomial distribution. If $p = 1/2$ and n is even, then $m = n/2$ is the unique median.

Covariance between Two Binomials

If two binomially distributed random variables X and Y are observed together, estimating their covariance can be useful. Using the definition of covariance, in the case $n = 1$ (thus being Bernoulli trials) we have

$$\text{Cov}(X,Y) = \text{E}(XY) - \mu_X \mu_Y.$$

The first term is non-zero only when both X and Y are one, and μ_X and μ_Y are equal to the two probabilities. Defining p_B as the probability of both happening at the same time, this gives

$$\text{Cov}(X,Y) = p_B - p_X p_Y,$$

and for n independent pairwise trials

$$\text{Cov}(X,Y)_n = n(p_B - p_X p_Y).$$

If X and Y are the same variable, this reduces to the variance formula given above.

Related Distributions

Sums of Binomials: If $X \sim \text{B}(n, p)$ and $Y \sim \text{B}(m, p)$ are independent binomial variables with the same probability p, then $X + Y$ is again a binomial variable; its distribution is

$$X + Y \sim B(n + m, p).$$

Conditional Binomials

If $X \sim \text{B}(n, p)$ and, conditional on X, $Y \sim \text{B}(X, q)$, then Y is a simple binomial variable with distribution

$$Y \sim B(n, pq).$$

For example imagine throwing n balls to a basket U_X and taking the balls that hit and throwing them to another basket U_Y. If p is the probability to hit U_X then $X \sim \text{B}(n, p)$ is the number of balls that hit U_X. If q is the probability to hit U_Y then the number of balls that hit U_Y is $Y \sim \text{B}(X, q)$ and therefore $Y \sim \text{B}(n, pq)$.

Bernoulli Distribution

The Bernoulli distribution is a special case of the binomial distribution, where $n = 1$. Symbolically, $X \sim \text{B}(1, p)$ has the same meaning as $X \sim \text{Bern}(p)$. Conversely, any binomial distribution, $\text{B}(n, p)$, is the distribution of the sum of n Bernoulli trials, $\text{Bern}(p)$, each with the same probability p.

Poisson Binomial Distribution

The binomial distribution is a special case of the Poisson binomial distribution, which is a sum of n independent non-identical Bernoulli trials $\text{Bern}(p_i)$. If X has the Poisson binomial distribution with $p_1 = \ldots = p_n = p$ then $X \sim \text{B}(n, p)$.

Normal Approximation

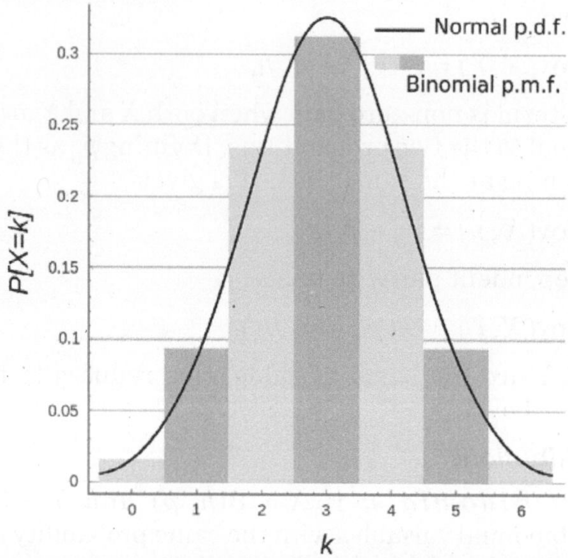

Figure: *Binomial probability mass function and normal probability density function approximation for n = 6 and p = 0.5*

If n is large enough, then the skew of the distribution is not too great. In this case a reasonable approximation to $B(n, p)$ is given by the normal distribution

$$\mathcal{N}(np, np(1-p)),$$

and this basic approximation can be improved in a simple way by using a suitable continuity correction. The basic approximation generally improves as n increases (at least 20) and is better when p is not near to 0 or 1. Various rules of thumb may be used to decide whether n is large enough, and p is far enough from the extremes of zero or one:

- One rule is that both $x=np$ and $n(1-p)$ must be greater than 5. However, the specific number varies from source to source, and depends on how good an approximation one wants; some sources give 10 which gives virtually the same results as the following rule for large n until n is very large (ex: $x=11$, $n=7752$).
- A second rule is that for $n > 5$ the normal approximation is adequate if

$$\left|\left(\frac{1}{\sqrt{n}}\right)\left(\sqrt{\frac{1-p}{p}} - \sqrt{\frac{p}{1-p}}\right)\right| < 0.3$$

- Another commonly used rule holds that the normal approximation is appropriate only if everything within 3 standard deviations of its mean is within the range of possible values, that is if

$$\mu \pm 3\sigma = np \pm 3\sqrt{np(1-p)} \in [0, n].$$

The following is an example of applying a continuity correction. Suppose one wishes to calculate $\Pr(X \leq 8)$ for a binomial random variable X. If Y has a distribution given by the normal approximation, then $\Pr(X \leq 8)$ is approximated by $\Pr(Y \leq 8.5)$. The addition of 0.5 is the continuity correction; the uncorrected normal approximation gives considerably less accurate results.

This approximation, known as de Moivre–Laplace theorem, is a huge time-saver when undertaking calculations by hand (exact calculations with large n are very onerous); historically, it was the first use of the normal distribution, introduced in Abraham de Moivre's book The Doctrine of Chances in 1738. Nowadays, it can be seen as a consequence of the central limit theorem since $B(n, p)$ is a sum of n independent, identically distributed Bernoulli variables with parameter p. This fact is the basis of a hypothesis test, a "proportion z-test", for the value of p using x/n, the sample proportion and estimator of p, in a common test statistic.

For example, suppose one randomly samples n people out of a large population and ask them whether they agree with a certain statement. The proportion of people who agree will of course depend on the sample. If groups of n people were sampled repeatedly and truly randomly, the proportions would follow an approximate normal distribution with mean equal to the true proportion p of agreement in the population and with standard deviation $\sigma = (p(1-p)/n)^{1/2}$. Large sample sizes n are good because the standard deviation, as a proportion of the expected value, gets smaller, which allows a more precise estimate of the unknown parameter p.

Poisson Approximation

The binomial distribution converges towards the Poisson distribution as the number of trials goes to infinity while the product np remains fixed. Therefore the Poisson distribution with parameter $\lambda = np$ can be used as an approximation to $B(n, p)$ of the binomial distribution if n is sufficiently large and p is sufficiently small. According to two rules of thumb, this approximation is good if $n \leq 20$ and $p \leq 0.05$, or if $n \leq 100$ and $np \leq 10$.

Limiting Distributions

- Poisson limit theorem: As n approaches – and p approaches 0 while np remains fixed at $\lambda > 0$ or at least np approaches $\lambda > 0$, then the Binomial(n, p) distribution approaches the Poisson distribution with expected value λ.

- de Moivre–Laplace theorem: As n approaches ∞ while p remains fixed, the distribution of

$$\frac{X-np}{\sqrt{np(1-p)}}$$

approaches the normal distribution with expected value 0 and variance 1. This result is sometimes loosely stated by saying that the distribution of X is asymptotically normal with expected value np and variance $np(1-p)$. This result is a specific case of the central limit theorem.

Beta Distribution

Beta distributions provide a family of conjugate prior probability distributions for binomial distributions in Bayesian inference. The domain of the beta distribution can be viewed as a probability, and in fact the beta distribution is often used to describe the distribution of a probability value p:

$$P(p;\alpha,\beta)=\frac{p^{\alpha-1}(1-p)^{\beta-1}}{\mathrm{B}(\alpha,\beta)}.$$

Confidence Intervals

Even for quite large values of n, the actual distribution of the mean is significantly nonnormal. Because of this problem several methods to estimate confidence intervals have been proposed.

Let n_1 be the number of successes out of n, the total number of trials, and let

$$\hat{p}=\frac{n_1}{n}$$

be the proportion of successes. Let $z_{\alpha/2}$ be the $100(1-\alpha/2)$th percentile of the standard normal distribution.

- Wald method

$$\hat{p}\pm z_{\frac{\alpha}{2}}\sqrt{\frac{\hat{p}(1-\hat{p})}{n}}.$$

A continuity correction of $0.5/n$ may be added.

- Agresti-Coull method

$$\tilde{p}\pm z_{\frac{\alpha}{2}}\sqrt{\frac{\tilde{p}(1-\tilde{p})}{n+z_{\frac{\alpha}{2}}^{2}}}.$$

Here the estimate of p is modified to

$$\tilde{p} = \frac{n_1 + \frac{1}{2} z^2_{\frac{\alpha}{2}}}{n + z^2_{\frac{\alpha}{2}}}$$

- ArcSine method

$$\sin^2\left(\arcsin\left(\sqrt{\hat{p}}\right) \pm \frac{z}{2\sqrt{n}}\right)$$

- Wilson (score) method

$$\frac{\hat{p} + \frac{1}{2n} z^2_{1-\frac{\alpha}{2}} \pm \frac{1}{2n} z_{1-\frac{\alpha}{2}} \sqrt{4n\hat{p}(1-\hat{p}) + z^2_{1-\frac{\alpha}{2}}}}{1 + \frac{1}{n} z^2_{1-\frac{\alpha}{2}}}.$$

The exact (Clopper-Pearson) method is the most conservative. The Wald method although commonly recommended in the text books is the most biased.[clarification needed]

Generating Binomial Random Variates

Methods for random number generation where the marginal distribution is a binomial distribution are well-established.

One way to generate random samples from a binomial distribution is to use an inversion algorithm. To do so, one must calculate the probability that P(X=k) for all values *k* from 0 through *n*. (These probabilities should sum to a value close to one, in order to encompass the entire sample space.) Then by using a Linear congruential generator to generate samples uniform between 0 and 1, one can transform the calculated samples U[0,1] into discrete numbers by using the probabilities calculated in step one.

Bounds for the Cumulative Distribution Function

For $k \leq np$, upper bounds for the lower tail of the distribution function can be derived. In particular, Hoeffding's inequality yields the bound

$$F(k;n,p) \leq \exp\left(-2\frac{(np-k)^2}{n}\right),$$

and Chernoff's inequality can be used to derive the bound

$$F(k;n,p) \leq \exp\left(-\frac{1}{2p}\frac{(np-k)^2}{n}\right).$$

Moreover, these bounds are reasonably tight when $p = 1/2$, since the following expression holds for all k e" $3n/8$

$$F(k;n,\tfrac{1}{2}) \geq \frac{1}{15}\exp\left(-\frac{16(\frac{n}{2}-k)^2}{n}\right).$$

However, the bounds do not work well for extreme values of p. In particular, as $p \to 1$, value $F(k;n,p)$ goes to zero (for fixed k, n with $k<n$) while the upper bound above goes to a positive constant. In this case a better bound is given by

$$F(k;n,p) \leq \exp\left(-nD\left(\frac{k}{n}\| \, |p\right)\right) \qquad \text{if } 0 < \frac{k}{n} < p$$

where $D(a \mid \mid p)$ is the relative entropy between an a-coin and a p-coin (i.e. between the Bernoulli(a) and Bernoulli(p) distribution):

$$D(a \parallel p) = (a)\log\frac{a}{p} + (1-a)\log\frac{1-a}{1-p}.$$

Asymptotically, this bound is reasonably tight. An equivalent formulation of the bound is

$$\Pr(X \geq k) = F(n-k;n,1-p) \leq \exp\left(-nD\left(\frac{k}{n}\| \, |p\right)\right) \qquad \text{if } p < \frac{k}{n} < 1.$$

Both these bounds are derived by directly minimising the Chernoff bound without any approximation in.

It can also be shown that,

$$\Pr(X \geq k) = F(n-k;n,1-p) \geq \frac{1}{(n+1)^2}\exp\left(-nD\left(\frac{k}{n}\| \, |p\right)\right) \qquad \text{if } p < \frac{k}{n} < 1.$$

This is proved using the method of types.

Chapter 8

Sampling-Theory of Estimation

In statistics, quality assurance, & survey methodology, sampling is concerned with the selection of a subset of individuals from within a statistical population to estimate characteristics of the whole population. Each observation measures one or more properties (such as weight, location, colour) of observable bodies distinguished as independent objects or individuals. In survey sampling, weights can be applied to the data to adjust for the sample design, particularly stratified sampling. Results from probability theory and statistical theory are employed to guide practice. In business and medical research, sampling is widely used for gathering information about a population.

The sampling process comprises several stages:

- Defining the population of concern
- Specifying a sampling frame, a set of items or events possible to measure
- Specifying a sampling method for selecting items or events from the frame
- Determining the sample size
- Implementing the sampling plan
- Sampling and data collecting
- Data which can be selected

Population Definition

Successful statistical practice is based on focused problem definition. In sampling, this includes defining the population from which our sample is drawn. A population can be defined as including all people or items with the characteristic one wishes to understand. Because there is very rarely enough time or money to gather information from everyone or everything in a population, the goal becomes finding a representative sample (or subset) of that population.

Sometimes what defines a population is obvious. For example, a manufacturer needs to decide whether a batch of material from production is of high enough quality to be released to the customer, or should be sentenced for scrap or rework due to poor quality. In this case, the batch is the population. Although the population of interest often consists of physical objects, sometimes we need to sample over time, space, or some combination of these dimensions. For instance, an investigation of supermarket staffing could examine checkout line length at various times, or a study on endangered penguins might aim to understand their usage of various hunting grounds over time. For the time dimension, the focus may be on periods or discrete occasions.

In other cases, our 'population' may be even less tangible. For example, Joseph Jagger studied the behaviour of roulette wheels at a casino in Monte Carlo, and used this to identify a biased wheel. In this case, the 'population' Jagger wanted to investigate was the overall behaviour of the wheel (i.e. the probability distribution of its results over infinitely many trials), while his 'sample' was formed from observed results from that wheel. Similar considerations arise when taking repeated measurements of some physical characteristic such as the electrical conductivity of copper.

This situation often arises when we seek knowledge about the cause system of which the *observed* population is an outcome. In such cases, sampling theory may treat the observed population as a sample from a larger 'superpopulation'. For example, a researcher might study the success rate of a new 'quit smoking' program on a test group of 100 patients, in order to predict the effects of the program if it were made available nationwide. Here the superpopulation is "everybody in the country, given access to this treatment" - a group which does not yet exist, since the program isn't yet available to all.

Note also that the population from which the sample is drawn may not be the same as the population about which we actually want information. Often there is large but not complete overlap between these two groups due to frame issues etc. Sometimes they may be entirely separate - for instance, we might study rats in order to get a better understanding of human health, or we might study records from people born in 2008 in order to make predictions about people born in 2009.

Time spent in making the sampled population and population of concern precise is often well spent, because it raises many issues, ambiguities and questions that would otherwise have been overlooked at this stage.

Sampling Frame

In the most straightforward case, such as the sentencing of a batch of material from production (acceptance sampling by lots), it is possible to identify and measure every single item in the population and to include any one of them in our sample. However, in the more general case this is not possible. There is no way to identify all rats in the set of all rats. Where voting is not compulsory, there is no way to identify which people will actually vote at a forthcoming election (in advance of the election). These imprecise populations are not amenable to sampling in any of the ways below and to which we could apply statistical theory.

As a remedy, we seek a sampling frame which has the property that we can identify every single element and include any in our sample. The most straightforward type of frame is a list of elements of the population (preferably the entire population) with appropriate contact information. For example, in an opinion poll, possible sampling frames include an electoral register and a telephone directory.

Probability and Nonprobability Sampling

Probability Sampling: A probability sample is a sample in which every unit in the population has a chance (greater than zero) of being selected in the sample, and this probability can be accurately determined. The combination of these traits makes it possible to produce unbiased estimates of population totals, by weighting sampled units according to their probability of selection.

Example: We want to estimate the total income of adults living in a given street. We visit each household in that street, identify all adults living there, and randomly select one adult from each household. (For example, we can allocate each person a random number, generated from a uniform distribution between 0 and 1, and select the person with the highest number in each household). We then interview the selected person and find their income.

People living on their own are certain to be selected, so we simply add their income to our estimate of the total. But a person living in a household of two adults has only a one-in-two chance of selection. To reflect this, when we come to such a household, we would count the selected person's income twice towards the total. (The person who is selected from that household can be loosely viewed as also representing the person who isn't selected.)

In the above example, not everybody has the same probability of selection; what makes it a probability sample is the fact that each person's probability is known. When every element in the population *does* have the same probability of selection, this is known as an 'equal probability of selection' (EPS) design. Such designs are also referred to as 'self-weighting' because all sampled units are given the same weight.

Probability sampling includes: Simple Random Sampling, Systematic Sampling, Stratified Sampling, Probability Proportional to Size Sampling, and Cluster or Multistage Sampling. These various ways of probability sampling have two things in common:

1. Every element has a known nonzero probability of being sampled and
2. involves random selection at some point.

Nonprobability Sampling

Nonprobability sampling is any sampling method where some elements of the population have *no* chance of selection (these are sometimes referred to as 'out of coverage'/'undercovered'), or where the probability of selection can't be accurately determined. It involves the selection of elements based on assumptions regarding the population of interest, which forms the criteria for selection. Hence, because the selection of elements is nonrandom, nonprobability sampling does not allow the estimation of sampling errors. These conditions give rise to exclusion bias, placing limits on how much information a sample can provide about the population. Information about the relationship between sample and population is limited, making it difficult to extrapolate from the sample to the population.

Example: We visit every household in a given street, and interview the first person to answer the door. In any household with more than one occupant, this is a nonprobability sample, because some people are more likely to answer the door (e.g. an unemployed person who spends most of their time at home is more likely to answer than an employed housemate who might be at work when the interviewer calls) and it's not practical to calculate these probabilities.

Nonprobability sampling methods include accidental sampling, quota sampling and purposive sampling. In addition, nonresponse effects may turn *any* probability design into a nonprobability design if the characteristics of nonresponse are not well understood, since nonresponse effectively modifies each element's probability of being sampled.

Sampling Methods

Within any of the types of frame identified above, a variety of sampling methods can be employed, individually or in combination. Factors commonly influencing the choice between these designs include:

- Nature and quality of the frame
- Availability of auxiliary information about units on the frame
- Accuracy requirements, and the need to measure accuracy
- Whether detailed analysis of the sample is expected
- Cost/operational concerns

Simple Random Sampling

In a simple random sample (SRS) of a given size, all such subsets of the frame are given an equal probability. Furthermore, any given *pair* of elements has the same chance of selection as any other such pair (and similarly for triples, and so on). This minimises bias and simplifies analysis of results. In particular, the variance between individual results within the sample is a good indicator of variance in the overall population, which makes it relatively easy to estimate the accuracy of results.

However, SRS can be vulnerable to sampling error because the randomness of the selection may result in a sample that doesn't reflect the makeup of the population. For instance, a simple random sample of ten people from a given country will *on average* produce five men and five women, but any given trial is likely to overrepresent one sex and underrepresent the other. (Systematic and stratified techniques), attempt to overcome this problem by "using information about the population" to choose a more "representative" sample.

SRS may also be cumbersome and tedious when sampling from an unusually large target population. In some cases, investigators are interested in "research questions specific" to subgroups of the population. For example, researchers might be interested in examining whether cognitive ability as a predictor of job performance is equally applicable across racial groups. SRS cannot accommodate the needs of researchers in this situation because it does not provide subsamples of the population. "Stratified sampling" addresses this weakness of SRS.

Systematic Sampling

Systematic sampling relies on arranging the study population according to some ordering scheme and then selecting elements at regular intervals through that ordered list. Systematic sampling involves a random start and then proceeds with the selection of every *k*th element from then onwards. In this case, *k*=(population size/sample size). It is important that the starting point is not automatically the first in the list, but is instead randomly chosen from within the first to the *k*th element in the list. A simple example would be to select every 10th name from the telephone directory (an 'every 10th' sample, also referred to as 'sampling with a skip of 10').

As long as the starting point is randomized, systematic sampling is a type of probability sampling. It is easy to implement and the stratification induced can make it efficient, *if* the variable by which the list is ordered is correlated with the variable of interest. 'Every 10th' sampling is especially useful for efficient sampling from databases.

For example, suppose we wish to sample people from a long street that starts in a poor area (house No. 1) and ends in an expensive district (house No. 1000). A simple random selection of addresses from this street could easily end up with too many from the high end and too few from the low end (or vice versa), leading to an unrepresentative sample. Selecting (e.g.) every 10th street number along the street ensures that the sample is spread evenly along the length of the street, representing all of these districts. (Note that if we always start at house #1 and end at #991, the sample is slightly biased towards the low end; by randomly selecting the start between #1 and #10, this bias is eliminated.

However, systematic sampling is especially vulnerable to periodicities in the list. If periodicity is present and the period is a multiple or factor of the interval used, the sample is especially likely to be *un*representative of the overall population, making the scheme less accurate than simple random sampling.

For example, consider a street where the odd-numbered houses are all on the north (expensive) side of the road, and the even-numbered houses are all on the south (cheap) side. Under the sampling scheme given above, it is impossible to get a representative sample; either the houses sampled will *all* be from the odd-numbered, expensive side, or they will *all* be from the even-numbered, cheap side.

Another drawback of systematic sampling is that even in scenarios where it is more accurate than SRS, its theoretical properties make it difficult to *quantify* that accuracy. (In the two examples of systematic sampling that are given above, much of the potential sampling error is due to variation between neighbouring houses - but because this method never selects two neighbouring houses, the sample will not give us any information on that variation.)

As described above, systematic sampling is an EPS method, because all elements have the same probability of selection (in the example given, one in ten). It is *not* 'simple random sampling' because different subsets of the same size have different selection probabilities - e.g. the set {4,14,24,...,994} has a one-in-ten probability of selection, but the set {4,13,24,34,...} has zero probability of selection.

Stratified Sampling

Where the population embraces a number of distinct categories, the frame can be organised by these categories into separate "strata." Each stratum is then sampled as an independent sub-population, out of which individual elements can be randomly selected. There are several potential benefits to stratified sampling.

First, dividing the population into distinct, independent strata can enable researchers to draw inferences about specific subgroups that may be lost in a more generalized random sample.

Second, utilizing a stratified sampling method can lead to more efficient statistical estimates (provided that strata are selected based upon relevance to the criterion in question, instead of availability of the samples). Even if a stratified sampling approach does not lead to increased statistical efficiency, such a tactic will not result in less efficiency than would simple random sampling, provided that each stratum is proportional to the group's size in the population.

Third, it is sometimes the case that data are more readily available for individual, pre-existing strata within a population than for the overall population; in such cases, using a stratified sampling approach may be more convenient than aggregating data across groups (though this may potentially be at odds with the previously noted importance of utilizing criterion-relevant strata).

Finally, since each stratum is treated as an independent population, different sampling approaches can be applied to different strata, potentially enabling researchers to use the approach best suited (or most cost-effective) for each identified subgroup within the population.

There are, however, some potential drawbacks to using stratified sampling. First, identifying strata and implementing such an approach can increase the cost and complexity of sample selection, as well as leading to increased complexity of population estimates. Second, when examining multiple criteria, stratifying variables may be related to some, but not to others, further complicating the design, and potentially reducing the utility of the strata. Finally, in some cases (such as designs with a large number of strata, or those with a specified minimum sample size per group), stratified sampling can potentially require a larger sample than would other methods (although in most cases, the required sample size would be no larger than would be required for simple random sampling.

A Stratified Sampling Approach is Most Effective When Three Conditions are Met

1. Variability within strata are minimized
2. Variability between strata are maximized
3. The variables upon which the population is stratified are strongly correlated with the desired dependent variable.

Advantages over Other Sampling Methods

1. Focuses on important subpopulations and ignores irrelevant ones.
2. Allows use of different sampling techniques for different subpopulations.
3. Improves the accuracy/efficiency of estimation.
4. Permits greater balancing of statistical power of tests of differences between strata by sampling equal numbers from strata varying widely in size.

Disadvantages

1. Requires selection of relevant stratification variables which can be difficult.
2. Is not useful when there are no homogeneous subgroups.
3. Can be expensive to implement.

Poststratification

Stratification is sometimes introduced after the sampling phase in a process called "poststratification". This approach is typically implemented due to a lack of prior knowledge of an appropriate stratifying variable or when the experimenter lacks the necessary information to create a stratifying variable during the sampling phase. Although the method is susceptible to the pitfalls of post hoc approaches, it can provide several benefits in the right situation. Implementation usually follows a simple random sample. In addition to allowing for stratification on an ancillary variable, poststratification can be used to implement weighting, which can improve the precision of a sample's estimates.

Oversampling

Choice-based sampling is one of the stratified sampling strategies. In choice-based sampling, the data are stratified on the target and a sample is taken from each stratum so that the rare target class will be more represented in the sample. The model is then built on this biased sample. The effects of the input variables on the target are often estimated with more precision with the choice-based sample even when a smaller overall sample size is taken, compared to a random sample. The results usually must be adjusted to correct for the oversampling.

Probability-Proportional-to-Size Sampling

In some cases the sample designer has access to an "auxiliary variable" or "size measure", believed to be correlated to the variable of interest, for each element in the population. These data can be used to improve accuracy in sample design. One option is to use the auxiliary variable as a basis for stratification, as discussed above.

Another option is probability proportional to size ('PPS') sampling, in which the selection probability for each element is set to be proportional to its size measure, up to a maximum of 1. In a simple PPS design, these selection probabilities can then be used as the basis for Poisson sampling. However, this has the drawback of variable sample size, and different portions of the population may still be over- or under-represented due to chance variation in selections.

Systematic sampling theory can be used to create a probability proportionate to size sample. This is done by treating each count within the size variable as a single sampling unit. Samples are then identified by selecting at even intervals among these counts within the size variable. This method is sometimes called PPS-sequential or monetary unit sampling in the case of audits or forensic sampling.

Example: Suppose we have six schools with populations of 150, 180, 200, 220, 260, and 490 students respectively (total 1500 students), and we want to use student population as the basis for a PPS sample of size three. To do this, we could allocate the first school numbers 1 to 150, the second school 151 to 330 (= 150 + 180), the third school 331 to 530, and so on to the last school (1011 to 1500). We then generate a random start between 1 and 500 (equal to 1500/3) and count through the school populations by multiples of 500. If our random start was 137, we would select the schools which have been allocated numbers 137, 637, and 1137, i.e. the first, fourth, and sixth schools.

The PPS approach can improve accuracy for a given sample size by concentrating sample on large elements that have the greatest impact on population estimates. PPS sampling is commonly used for surveys of businesses, where element size varies greatly and auxiliary information is often available—for instance, a survey attempting to measure the number of guest-nights spent in hotels might use each hotel's number of rooms as an auxiliary variable. In some cases, an older measurement of the variable of interest can be used as an auxiliary variable when attempting to produce more current estimates.

Cluster Sampling

Sometimes it is more cost-effective to select respondents in groups ('clusters'). Sampling is often clustered by geography, or by time periods. (Nearly all samples are in some sense 'clustered' in time - although this is rarely taken into account in the analysis.) For instance, if surveying households within a city, we might choose to select 100 city blocks and then interview every household within the selected blocks.

Clustering can reduce travel and administrative costs. In the example above, an interviewer can make a single trip to visit several households in one block, rather than having to drive to a different block for each household. It also means that one does not need a sampling frame listing all elements in the target population. Instead, clusters can be chosen from a cluster-level frame, with an element-level frame created only for the selected clusters. In the example above, the sample only requires a block-level city map for initial selections, and then a household-level map of the 100 selected blocks, rather than a household-level map of the whole city.

Cluster sampling generally increases the variability of sample estimates above that of simple random sampling, depending on how the clusters differ between themselves, as compared with the within-cluster variation. For this reason, cluster sampling requires a larger sample than SRS to achieve the same level of accuracy - but cost savings from clustering might still make this a cheaper option.

Cluster sampling is commonly implemented as multistage sampling. This is a complex form of cluster sampling in which two or more levels of units are embedded one in the other. The first stage consists of constructing the clusters that will be used to sample from. In the second stage, a sample of primary units is randomly selected from each cluster (rather than using all units contained in all selected clusters). In following stages, in each of those selected clusters, additional samples of units are selected, and so on. All ultimate units (individuals, for instance) selected at the last step of this procedure are then surveyed. This technique, thus, is essentially the process of taking random subsamples of preceding random samples.

Multistage sampling can substantially reduce sampling costs, where the complete population list would need to be constructed (before other sampling methods could be applied). By eliminating the work involved in describing clusters that are not selected, multistage sampling can reduce the large costs associated with traditional cluster sampling. However, each sample may not be a full representative of the whole population.

Quota Sampling

In quota sampling, the population is first segmented into mutually exclusive sub-groups, just as in stratified sampling. Then judgement is used to select the subjects or units from each segment based on a specified proportion. For example, an interviewer may be told to sample 200 females and 300 males between the age of 45 and 60.

It is this second step which makes the technique one of non-probability sampling. In quota sampling the selection of the sample is non-random. For example interviewers might be tempted to interview those who look most helpful. The problem is that these samples may be biased because not everyone gets a chance of selection. This random element is its greatest weakness and quota versus probability has been a matter of controversy for several years.

Minimax Sampling

In imbalanced datasets, where the sampling ratio does not follow the population statistics, one can resample the dataset in a conservative manner called minimax sampling. The minimax sampling has its origin in Anderson minimax ratio whose value is proved to be 0.5: in a binary classification, the class-sample sizes should be chosen equally. This ratio can be proved to be minimax ratio only under the assumption of LDA classifier with Gaussian distributions. The notion of minimax sampling is recently developed for a general class of classification rules, called class-wise smart classifiers. In this case, the sampling ratio of classes is selected so that the worst case classifier error over all the possible population statistics for class prior probabilities, would be the best.

Accidental Sampling

Accidental sampling (sometimes known as grab, convenience or opportunity sampling) is a type of nonprobability sampling which involves the sample being drawn from that part of the population which is close to hand. That is, a population is selected because it is readily available and convenient. It may be through meeting the person or including a person in the sample when one meets them or chosen by finding them through technological means such as the internet or through phone. The researcher using such a sample cannot scientifically make generalizations about the total population from this sample because it would not be representative enough. For example, if the interviewer were to conduct such a survey at a shopping center early in the morning on a given day, the people that he/she could interview would be limited to those given there at that given time, which would not represent the views of other members of society in such an area, if the survey were to be conducted at different times of day and several times per week. This type of sampling is most useful for pilot testing. Several important considerations for researchers using convenience samples include:

1. Are there controls within the research design or experiment which can serve to lessen the impact of a non-random convenience sample, thereby ensuring the results will be more representative of the population?
2. Is there good reason to believe that a particular convenience sample would or should respond or behave differently than a random sample from the same population?
3. Is the question being asked by the research one that can adequately be answered using a convenience sample?

In social science research, snowball sampling is a similar technique, where existing study subjects are used to recruit more subjects into the sample. Some variants of snowball sampling, such as respondent driven sampling, allow calculation of selection probabilities and are probability sampling methods under certain conditions.

Line-Intercept Sampling

Line-intercept sampling is a method of sampling elements in a region whereby an element is sampled if a chosen line segment, called a "transect", intersects the element.

Panel Sampling

Panel sampling is the method of first selecting a group of participants through a random sampling method and then asking that group for (potentially the same) information several times over a period of time. Therefore, each participant is interviewed at two or more time points; each period of data collection is called a "wave". The method was developed by sociologist Paul Lazarsfeld in 1938 as a means of studying political campaigns. This longitudinal sampling-method allows estimates of changes in the population, for example with regard to chronic illness to job stress to weekly food expenditures. Panel sampling can also be used to inform researchers about within-person health changes due to age or to help explain changes in continuous dependent variables such as spousal interaction. There have been several proposed methods of analyzing panel data, including MANOVA, growth curves, and structural equation modelling with lagged effects.

Replacement of Selected Units

Sampling schemes may be *without replacement* ('WOR'—no element can be selected more than once in the same sample) or *with replacement* ('WR'—an element may appear multiple times in the one sample). For example, if we catch fish, measure them, and immediately return them to the water before continuing with the sample, this is a WR design, because we might end up catching and measuring the same fish more than once. However, if we do not return the fish to the water (e.g., if we eat the fish), this becomes a WOR design.

Sample Size

Formulas, tables, and power function charts are well known approaches to determine sample size.

Steps for Using Sample Size Tables

1. Postulate the effect size of interest, α, and β.
2. Check sample size table
   1. Select the table corresponding to the selected α
   2. Locate the row corresponding to the desired power
   3. Locate the column corresponding to the estimated effect size.
   4. The intersection of the column and row is the minimum sample size required.

Sampling and Data Collection

Good data collection involves:

- Following the defined sampling process
- Keeping the data in time order
- Noting comments and other contextual events
- Recording non-responses

Errors in Sample Surveys

Survey results are typically subject to some error. Total errors can be classified into sampling errors and non-sampling errors. The term "error" here includes systematic biases as well as random errors.

Sampling Errors and Biases

Sampling errors and biases are induced by the sample design. They include:

1. Selection bias: When the true selection probabilities differ from those assumed in calculating the results.
2. Random sampling error: Random variation in the results due to the elements in the sample being selected at random.

Non-Sampling Error

Non-sampling errors are other errors which can impact the final survey estimates, caused by problems in data collection, processing, or sample design. They include:

1. Overcoverage: Inclusion of data from outside of the population.
2. Undercoverage: Sampling frame does not include elements in the population.
3. Measurement error: e.g. when respondents misunderstand a question, or find it difficult to answer.
4. Processing error: Mistakes in data coding.
5. Non-response: Failure to obtain complete data from all selected individuals.

After sampling, a review should be held of the exact process followed in sampling, rather than that intended, in order to study any effects that any divergences might have on subsequent analysis. A particular problem is that of *non-response.*

Two major types of nonresponse exist: unit nonresponse (referring to lack of completion of any part of the survey) and item nonresponse (submission or participation in survey but failing to complete one or more components/questions of the survey). In survey sampling, many of the individuals identified as part of the sample may be unwilling to participate, not have the time to participate (opportunity cost), or survey administrators may not have been able to contact them. In this case, there is a risk of differences, between respondents and nonrespondents, leading to biased estimates of population parameters. This is often addressed by improving survey design, offering incentives, and conducting follow-up studies which make a repeated attempt to contact the unresponsive and to characterize their similarities and differences with the rest of the frame. The effects can also be mitigated by weighting the data when population benchmarks are available or by imputing data based on answers to other questions.

Nonresponse is particularly a problem in internet sampling. Reasons for this problem include improperly designed surveys, over-surveying (or survey fatigue), and the fact that potential participants hold multiple e-mail addresses, which they don't use anymore or don't check regularly.

Survey Weights

In many situations the sample fraction may be varied by stratum and data will have to be weighted to correctly represent the population.

Thus for example, a simple random sample of individuals in the United Kingdom might include some in remote Scottish islands who would be inordinately expensive to sample. A cheaper method would be to use a stratified sample with urban and rural strata. The rural sample could be under-represented in the sample, but weighted up appropriately in the analysis to compensate.

More generally, data should usually be weighted if the sample design does not give each individual an equal chance of being selected. For instance, when households have equal selection probabilities but one person is interviewed from within each household, this gives people from large households a smaller chance of being interviewed. This can be accounted for using survey weights. Similarly, households with more than one telephone line have a greater chance of being selected in a random digit dialing sample, and weights can adjust for this.

Weights can also serve other purposes, such as helping to correct for non-response.

Expected Value by Mathematical Expectation

The process of placing classified data into tabular form is known as tabulation. A table is a symmetric arrangement of statistical data in rows and columns. Rows are horizontal arrangements whereas columns are vertical arrangements. It may be simple, double or complex depending upon the type of classification.

In probability theory, the expected value of a random variable is intuitively the long-run average value of repetitions of the experiment it represents. For example, the expected value of a die roll is 3.5 because, roughly speaking, the average of an extremely large number of dice rolls is practically always nearly equal to 3.5. Less roughly, the law of large numbers guarantees that the arithmetic mean of the values almost surely converges to the expected value as the number of repetitions goes to infinity. The expected value is also known as the expectation, mathematical expectation, EV, mean, or first moment.

More practically, the expected value of a discrete random variable is the probability-weighted average of all possible values. In other words, each possible value the random variable can assume is multiplied by its probability of occurring, and the resulting products are summed to produce the expected value. The same works for continuous random variables, except the sum is replaced by an integral and the probabilities by probability densities. The formal definition subsumes both of these and also works for distributions which are neither discrete nor continuous: the expected value of a random variable is the integral of the random variable with respect to its probability measure.

The expected value does not exist for random variables having some distributions with large "tails", such as the Cauchy distribution. For random variables such as these, the long-tails of the distribution prevent the sum/integral from converging.

The expected value is a key aspect of how one characterizes a probability distribution; it is one type of location parameter. By contrast, the variance is a measure of dispersion of the possible values of the random variable around the expected value. The variance itself is defined in terms of two expectations: it is the expected value of the squared deviation of the variable's value from the variable's expected value.

The expected value plays important roles in a variety of contexts. In regression analysis, one desires a formula in terms of observed data that will give a "good" estimate of the parameter giving the effect of some explanatory variable upon a dependent variable. The formula will give different estimates using different samples of data, so the estimate it gives is itself a random variable. A formula is typically considered good in this context if it is an unbiased estimator—that is, if the expected value of the estimate (the average value it would give over an arbitrarily large number of separate samples) can be shown to equal the true value of the desired parameter.

In decision theory, and in particular in choice under uncertainty, an agent is described as making an optimal choice in the context of incomplete information. For risk neutral agents, the choice involves using the expected values of uncertain quantities, while for risk averse agents it involves maximizing the expected value of some objective function such as a von Neumann-Morgenstern utility function.

Definition

Univariate Discrete Random Variable, Finite Case: Suppose random variable X can take value x_1 with probability p_1, value x_2 with probability p_2, and so on, up to value x_k with probability p_k. Then the expectation of this random variable X is defined as

$$\mathrm{E}[X] = x_1 p_1 + x_2 p_2 + \cdots + x_k p_k \,.$$

Since all probabilities p_i add up to one ($p_1 + p_2 + \ldots + p_k = 1$), the expected value can be viewed as the weighted average, with p_i's being the weights:

$$\mathrm{E}[X] = \frac{x_1 p_1 + x_2 p_2 + \cdots + x_k p_k}{1} = \frac{x_1 p_1 + x_2 p_2 + \cdots + x_k p_k}{p_1 + p_2 + \cdots + p_k}.$$

If all outcomes x_i are equally likely (that is, $p_1 = p_2 = \ldots = p_k$), then the weighted average turns into the simple average. This is intuitive:

the expected value of a random variable is the average of all values it can take; thus the expected value is what one expects to happen *on average*. If the outcomes x_i are not equally probable, then the simple average must be replaced with the weighted average, which takes into account the fact that some outcomes are more likely than the others. The intuition however remains the same: the expected value of X is what one expects to happen *on average*.

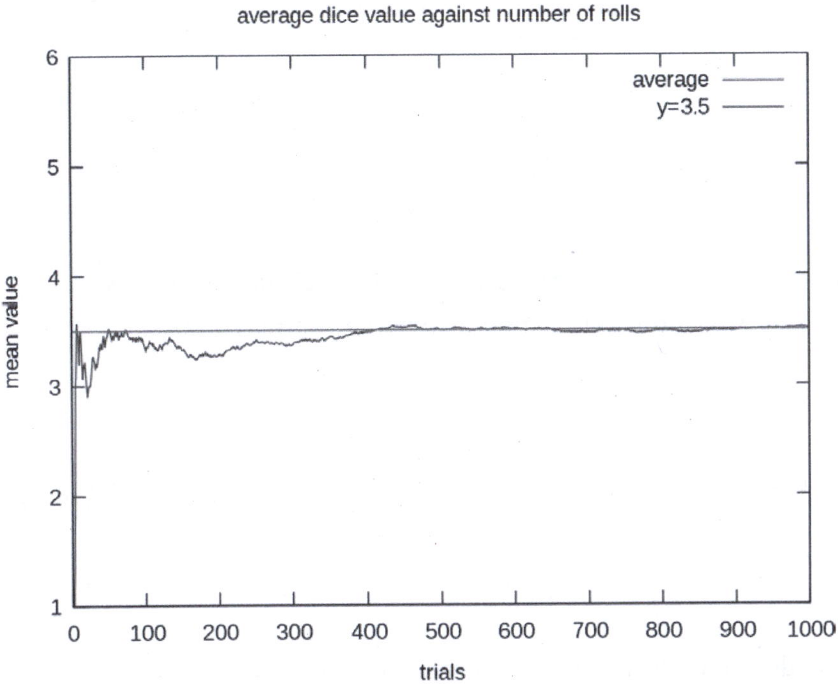

Figure: *An illustration of the convergence of sequence averages of rolls of a die to the expected value of 3.5 as the number of rolls (trials) grows.*

Example 1: Let X represent the outcome of a roll of a fair six-sided die. More specifically, X will be the number of pips showing on the top face of the die after the toss. The possible values for X are 1, 2, 3, 4, 5, and 6, all equally likely (each having the probability of 1/6). The expectation of X is

$$\mathrm{E}[X] = 1\cdot\frac{1}{6} + 2\cdot\frac{1}{6} + 3\cdot\frac{1}{6} + 4\cdot\frac{1}{6} + 5\cdot\frac{1}{6} + 6\cdot\frac{1}{6} = 3.5.$$

If one rolls the die n times and computes the average (arithmetic mean) of the results, then as n grows, the average will almost surely converge to the expected value, a fact known as the strong law of large numbers. One example sequence of ten rolls of the die is 2, 3, 1, 2, 5, 6, 2, 2, 2, 6, which has the average of 3.1, with the distance of 0.4 from the expected value of 3.5. The convergence is relatively slow: the probability that the average falls within the range 3.5 ± 0.1 is 21.6% for ten rolls, 46.1% for a hundred rolls and 93.7% for a thousand rolls. More generally, the rate of convergence can be roughly quantified by e.g. Chebyshev's inequality and the Berry-Esseen theorem.

Example 2. The roulette game consists of a small ball and a wheel with 38 numbered pockets around the edge. As the wheel is spun, the ball bounces around randomly until it settles down in one of the pockets. Suppose random variable X represents the (monetary) outcome of a \$1 bet on a single number ("straight up" bet). If the bet wins (which happens with probability 1/38), the payoff is \$35; otherwise the player loses the bet. The expected profit from such a bet will be

$$\text{E[gain from \$1 bet]} = -\$1 \cdot \frac{37}{38} + \$35 \cdot \frac{1}{38} = -\$0.0526.$$

Univariate Discrete Random Variable, Countable Case

Let X be a discrete random variable taking values x

1, x

2, ... with probabilities p

1, p

2, ... respectively. Then the expected value of this random variable is the infinite sum

$$\mathrm{E}[X] = \sum_{i=1}^{\infty} x_i p_i,$$

provided that this series converges absolutely (that is, the sum must remain finite if we were to replace all x

i's with their absolute values). If this series does not converge absolutely, we say that the expected value of X does not exist.

For example, suppose random variable X takes values 1, –2, 3, –4, ..., with respective probabilities $c/1^2$, $c/2^2$, $c/3^2$, $c/4^2$, ..., where $c = 6/\pi^2$ is a normalizing constant that ensures the probabilities sum up to one. Then the infinite sum

$$\sum_{i=1}^{\infty} x_i p_i = c\left(1 - \frac{1}{2} + \frac{1}{3} - \frac{1}{4} + \cdots\right)$$

converges and its sum is equal to ln(2) C≃ 0.69315. However it would be incorrect to claim that the expected value of X is equal to this number—in fact E[X] does not exist, as this series does not converge absolutely.

Univariate Continuous Random Variable

If the probability distribution of X admits a probability density function $f(x)$, then the expected value can be computed as

$$\mathrm{E}[X] = \int_{-\infty}^{\infty} x f(x)\mathrm{d}x.$$

General Definition

In general, if X is a random variable defined on a probability space (Ω, Σ, P), then the expected value of X, denoted by E[X],)#X*#, X or E[X], is defined as the Lebesgue integral

$$\mathrm{E}[X] = \int_{\Omega} X \mathrm{d}P = \int_{\Omega} X(\omega)P(\mathrm{d}\omega)$$

When this integral exists, it is defined as the expectation of X. Note that not all random variables have a finite expected value, since the integral may not converge absolutely; furthermore, for some it is not defined at all (e.g., Cauchy distribution). Two variables with the same probability distribution will have the same expected value, if it is defined.

It follows directly from the discrete case definition that if X is a constant random variable, i.e. $X = b$ for some fixed real number b, then the expected value of X is also b.

The expected value of a measurable function of X, $g(X)$, given that X has a probability density function $f(x)$, is given by the inner product of f and g:

$$\mathrm{E}[g(X)] = \int_{-\infty}^{\infty} g(x)f(x)\mathrm{d}x.$$

This is sometimes called the law of the unconscious statistician. Using representations as Riemann–Stieltjes integral and integration by parts the formula can be restated as

$$\mathrm{E}[g(X)] = \int_{a}^{\infty} g(x)\mathrm{dP}(X \le x) = \begin{cases} g(a) + \int_{a}^{\infty} g'(x)\mathrm{P}(X > x)\mathrm{d}x & \text{if } \mathrm{P}(g(X) \ge g(a)) = 1 \\ g(b) - \int_{-\infty}^{b} g'(x)\mathrm{P}(X \le x)\mathrm{d}x & \text{if } \mathrm{P}(g(X) \le g(b)) = 1. \end{cases}$$

As a special case let α denote a positive real number. Then

$$\mathrm{E}\left[|X|^{\alpha}\right] = \alpha \int_{0}^{\infty} t^{\alpha-1}\mathrm{P}(|X| > t)\mathrm{d}t.$$

In particular, if $\alpha = 1$ and $\Pr[X \ge 0] = 1$, then this reduces to

$$\mathrm{E}[|X|] = \mathrm{E}[X] = \int_{0}^{\infty} \{1 - F(t)\}\mathrm{d}t,$$

where F is the cumulative distribution function of X. This last identity is an instance of what, in a non-probabilistic setting, has been called the layer cake representation.

The law of the unconscious statistician applies also to a measurable function g of several random variables X_1, ... X_n having a joint density f:

$$\mathrm{E}[g(X_1,\ldots,X_n)] = \int_{-\infty}^{\infty}\cdots\int_{-\infty}^{\infty} g(x_1,\cdots,x_n)\,f(x_1,\cdots,x_n)\,\mathrm{d}x_1\cdots\mathrm{d}x_n.$$

Properties

Constants: The expected value of a constant is equal to the constant itself; i.e., if c is a constant, then $\mathrm{E}[c] = c$.

Monotonicity

If X and Y are random variables such that $X \le Y$ almost surely, then $\mathrm{E}[X] \le \mathrm{E}[Y]$.

Linearity

The expected value operator (or expectation operator) E is linear in the sense that

$$\begin{aligned} \mathrm{E}[X+c] &= \mathrm{E}[X]+c \\ \mathrm{E}[X+Y] &= \mathrm{E}[X]+\mathrm{E}[Y] \\ \mathrm{E}[aX] &= a\,\mathrm{E}[X] \end{aligned}$$

Note that the second result is valid even if X is not statistically independent of Y. Combining the results from previous three equations, we can see that

$$\mathrm{E}[aX+bY+c] = a\,\mathrm{E}[X]+b\,\mathrm{E}[Y]+c$$

for any two random variables X and Y (which need to be defined on the same probability space) and any real numbers a, b and c.

Iterated Expectation

Iterated Expectation for Discrete Random Variables: For any two discrete random variables X, Y one may define the conditional expectation:

$$\mathrm{E}[X \mid Y=y] = \sum_x x\cdot\mathrm{P}(X=x \mid Y=y).$$

which means that $\mathrm{E}[X \mid Y=y]$ is a function of y. Let $g(y)$ be that function of y; then the notation $\mathrm{E}[X \mid Y]$ is then a random variable in its own right, equal to $g(Y)$.

Lemma. Then the expectation of X satisfies:

$$\mathrm{E}[X] = \mathrm{E}\big[\mathrm{E}[X \mid Y]\big].$$

Proof.:

$$\begin{aligned}
\mathrm{E}\big[\mathrm{E}[X \mid Y]\big] &= \sum_y \mathrm{E}[X \mid Y = y]\cdot \mathrm{P}(Y = y) \\
&= \sum_y \left(\sum_x x\cdot \mathrm{P}(X = x \mid Y = y) \right)\cdot \mathrm{P}(Y = y) \\
&= \sum_y \sum_x x\cdot \mathrm{P}(X = x \mid Y = y)\cdot \mathrm{P}(Y = y) \\
&= \sum_y \sum_x x\cdot \mathrm{P}(Y = y \mid X = x)\cdot \mathrm{P}(X = x) \\
&= \sum_x x\cdot \mathrm{P}(X = x)\cdot \left(\sum_y \mathrm{P}(Y = y \mid X = x) \right) \\
&= \sum_x x\cdot \mathrm{P}(X = x) \\
&= \mathrm{E}[X]
\end{aligned}$$

The left-hand side of this equation is referred to as the *iterated expectation*. The equation is sometimes called the *tower rule* or the *tower property*; it is treated under law of total expectation.

Iterated Expectation for Continuous Random Variables

In the continuous case, the results are completely analogous. The definition of conditional expectation would use inequalities, density functions, and integrals to replace equalities, mass functions, and summations, respectively. However, the main result still holds:

$$\mathrm{E}[X] = \mathrm{E}[\mathrm{E}[X \mid Y]]$$

Inequality

If a random variable X is always less than or equal to another random variable Y, the expectation of X is less than or equal to that of Y:

$$\text{If } X \le Y, \text{ then } \mathrm{E}[X] \le \mathrm{E}[Y].$$

In particular, if we set Y to $|X|$ we know $X \le Y$ and $-X \le Y$. Therefore we know $\mathrm{E}[X] \le \mathrm{E}[Y]$ and $\mathrm{E}[-X] \le \mathrm{E}[Y]$. From the linearity of expectation we know $-\mathrm{E}[X] \le \mathrm{E}[Y]$. Therefore the absolute value of expectation of a random variable is less than or equal to the expectation of its absolute value:

$$|\mathrm{E}[X]| \le \mathrm{E}[|X|]$$

Non-Multiplicativity

If one considers the joint probability density function of X and Y, say $j(x,y)$, then the expectation of XY is

$$E[XY] = \iint xy\, j(x,y)\,dx\,dy.$$

In general, the expected value operator is not multiplicative, i.e. $E[XY]$ is not necessarily equal to $E[X]\cdot E[Y]$. In fact, the amount by which multiplicativity fails is called the covariance:

$$\text{Cov}(X,Y) = E[XY] - E[X]E[Y].$$

Thus multiplicativity holds precisely when $\text{Cov}(X, Y) = 0$, in which case X and Y are said to be uncorrelated (independent variables are a notable case of uncorrelated variables).

Now if X and Y are independent, then by definition $j(x,y) = f(x)g(y)$ where f and g are the marginal PDFs for X and Y. Then

$$\begin{aligned} E[XY] &= \iint xy\, j(x,y)\,dx\,dy = \iint xy f(x)g(y)\,dy\,dx \\ &= \left[\int x f(x)\,dx\right]\left[\int y g(y)\,dy\right] = E[X]E[Y] \end{aligned}$$

and $\text{Cov}(X, Y) = 0$.

Observe that independence of X and Y is required only to write $j(x, y) = f(x)g(y)$, and this is required to establish the second equality above. The third equality follows from a basic application of the Fubini-Tonelli theorem.

Functional Non-Invariance

In general, the expectation operator and functions of random variables do not commute; that is

$$E[g(X)] = \int_\Omega g(X)\,dP \neq g(E[X]),$$

A notable inequality concerning this topic is Jensen's inequality, involving expected values of convex (or concave) functions.

Uses and Applications

It is possible to construct an expected value equal to the probability of an event by taking the expectation of an indicator function that is one if the event has occurred and zero otherwise. This relationship can be used to translate properties of expected values into properties of probabilities, e.g. using the law of large numbers to justify estimating probabilities by frequencies.

The expected values of the powers of X are called the moments of X; the moments about the mean of X are expected values of powers of

If X is a purely discrete random variable, then it attains values x_1, x_2, ... with probability $p_i = \mathrm{P}(x_i)$, and the CDF of X will be discontinuous at the points x_i and constant in between:

$$F(x) = \mathrm{P}(X \le x) = \sum_{x_i \le x} \mathrm{P}(X = x_i) = \sum_{x_i \le x} p(x_i).$$

If the CDF F of a real valued random variable X is continuous, then X is a continuous random variable; if furthermore F is absolutely continuous, then there exists a Lebesgue-integrable function $f(x)$ such that

$$F(b) - F(a) = \mathrm{P}(a < X \le b) = \int_a^b f(x)dx$$

for all real numbers a and b. The function f is equal to the derivative of F almost everywhere, and it is called the probability density function of the distribution of X.

Examples

As an example, suppose X is uniformly distributed on the unit interval [0, 1]. Then the CDF of X is given by

$$F(x) = \begin{cases} 0 & : x < 0 \\ x & : 0 \le x < 1 \\ 1 & : 1 \le x. \end{cases}$$

Suppose instead that X takes only the discrete values 0 and 1, with equal probability. Then the CDF of X is given by

$$F(x) = \begin{cases} 0 & : x < 0 \\ 1/2 & : 0 \le x < 1 \\ 1 & : 1 \le x. \end{cases}$$

Derived Functions

Complementary Cumulative Distribution Function (Tail Distribution): Sometimes, it is useful to study the opposite question and ask how often the random variable is *above* a particular level. This is called the complementary cumulative distribution function (ccdf) or simply the tail distribution or exceedance, and is defined as

$$\bar{F}(x) = \mathrm{P}(X > x) = 1 - F(x).$$

This has applications in statistical hypothesis testing, for example, because the one-sided p-value is the probability of observing a test statistic *at least* as extreme as the one observed. Thus, provided that the test statistic, T, has a continuous distribution, the one-sided p-value is simply given by the ccdf: for an observed value t of the test statistic

$$p = \mathrm{P}(T \geq t) = \mathrm{P}(T > t) = 1 - F_T(t).$$

In survival analysis, $\bar{F}(x)$ is called the survival function and denoted $S(x)$, while the term *reliability function* is common in engineering.

Properties

- For a non-negative continuous random variable having an expectation, Markov's inequality states that

$$\bar{F}(x) \leq \frac{\mathbb{E}(X)}{x}.$$

- As $x \to \infty, \bar{F}(x) \to 0$, and in fact $\bar{F}(x) = o(1/x)$ provided that $\mathbb{E}(X)$ is finite.

Proof: Assuming X has a density function f, for any $c > 0$

$$\mathbb{E}(X) = \int_0^\infty xf(x)dx \geq \int_0^c xf(x)dx + c\int_c^\infty f(x)dx$$

Then, on recognising $\bar{F}(c) = \int_c^\infty f(x)dx$ and rearranging terms,

$$0 \leq c\bar{F}(c) \leq \mathbb{E}(X) - \int_0^c xf(x)dx \to 0 \text{ as } c \to \infty$$

as claimed.

Folded Cumulative Distribution

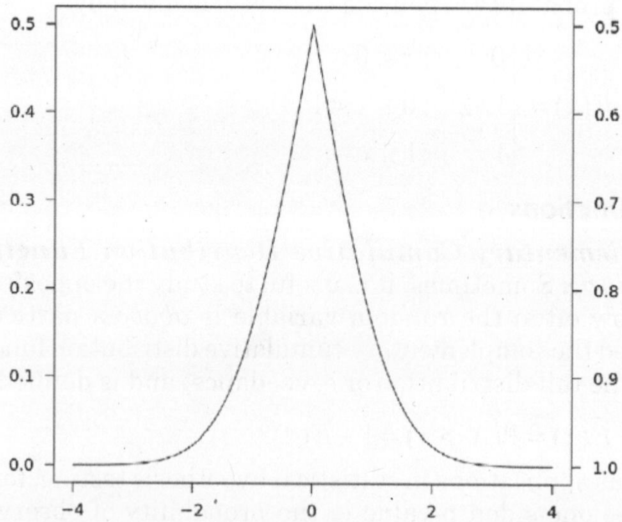

Figure: *Example of the folded cumulative distribution for a normal distribution function with an expected value of 0 and a standard deviation of 1.*

While the plot of a cumulative distribution often has an S-like shape, an alternative illustration is the folded cumulative distribution or mountain plot, which folds the top half of the graph over, thus using two scales, one for the upslope and another for the downslope. This form of illustration emphasises the median and dispersion (the mean absolute deviation from the median) of the distribution or of the empirical results.

Inverse Distribution Function (Quantile Function)

If the CDF F is strictly increasing and continuous then $F^{-1}(y), y \in [0,1]$, is the unique real number $x$such that $F(x) = y$. In such a case, this defines the inverse distribution function or quantile function.

Unfortunately, the distribution does not, in general, have an inverse. One may define, for $y \in [0,1]$, the generalized inverse distribution function:

$$F^{-1}(y) = \inf\{x \in \mathbb{R} : F(x) \geq y\}.$$

- Example 1: The median is $F^{-1}(0.5)$.
- Example 2: Put $\tau = F^{-1}(0.95)$. Then we call τ the 95th percentile.

The inverse of the cdf is called the quantile function.

The inverse of the cdf can be used to translate results obtained for the uniform distribution to other distributions. Some useful properties of the inverse cdf are:

1. F^{-1} is nondecreasing
2. $F^{-1}(F(x)) \leq x$
3. $F(F^{-1}(y)) \geq y$
4. $F^{-1}(y) \leq x$ if and only if $y \leq F(x)$
5. If Y has a $U[0,1]$ distribution then $F^{-1}(Y)$ is distributed as F. This is used in random number generation using the inverse transform sampling-method.
6. If $\{X_\alpha\}$ is a collection of independent F-distributed random variables defined on the same sample space, then there exist random variables Y_α such that Y_α is distributed as $U[0,1]$ and $F^{-1}(Y_\alpha) = X_\alpha$ with probability 1 for all α.

Multivariate Case

When dealing simultaneously with more than one random variable the *joint* cumulative distribution function can also be defined. For example, for a pair of random variables X, Y, the joint CDF F is given by

$$F(x, y) = \mathrm{P}(X \le x, Y \le y),$$

where the right-hand side represents the probability that the random variable X takes on a value less than or equal to x and that Y takes on a value less than or equal to y.

Every Multivariate CDF is

1. Monotonically non-decreasing for each of its variables
2. Right-continuous for each of its variables.
3. $0 \le F(x_1, \ldots, x_n) \le 1$
4. $\lim_{x_1, \ldots, x_n \to +\infty} F(x_1, \ldots, x_n) = 1$ and $\lim_{x_i \to -\infty} F(x_1, \ldots, x_n) = 0,$ for all i

Use in Statistical Analysis

The concept of the cumulative distribution function makes an explicit appearance in statistical analysis in two (similar) ways. Cumulative frequency analysis is the analysis of the frequency of occurrence of values of a phenomenon less than a reference value. The empirical distribution function is a formal direct estimate of the cumulative distribution function for which simple statistical properties can be derived and which can form the basis of various statistical hypothesis tests. Such tests can assess whether there is evidence against a sample of data having arisen from a given distribution, or evidence against two samples of data having arisen from the same (unknown) population distribution.

Kolmogorov–Smirnov and Kuiper's Tests

The Kolmogorov–Smirnov test is based on cumulative distribution functions and can be used to test to see whether two empirical distributions are different or whether an empirical distribution is different from an ideal distribution. The closely related Kuiper's test is useful if the domain of the distribution is cyclic as in day of the week. For instance Kuiper's test might be used to see if the number of tornadoes varies during the year or if sales of a product vary by day of the week or day of the month.

Bibliography

Abramowitz, Milton; Stegun, Irene A.: *Handbook of Mathematical Functions with Formulas, Graphs, and Mathematical Tables,* Dover Publications, New York, 1972.

Allen, Elizabeth; Triantaphillidou, Sophie: *The Manual of Photography*, Taylor & Francis, NJ, 2011.

Ambartzumian, R. V.: *Factorization Calculus and Geometric Probability*, Cambridge University Press, NY, 1990.

Angel, Allen R.; Porter, Stuart R.: *A Survey of Mathematics with Applications,* Addison-Wesley, NY, 1989.

Apostol, T.M.: *NIST Handbook of Mathematical Functions*, Cambridge University Press, NY, 2010.

Bernstein, Stephen; Bernstein, Ruth: *Schaum's Cutline of Theory and Problems of Elements of Statistics. I, Descriptive statistics and probability*, McGraw-Hill, New York, 1999.

Bourbaki, Nicolas: *General Topology,* Springer-Verlag, Berlin, New York, 1998.

Cameron, Peter J.: *Combinatorics: Topics, Techniques, Algorithms*, Cambridge University Press, Cambridge, 1994.

D.E. Smith: *History of Mathematics,* Dover, NJ, 1958.

Daren Bieniek, Randy Dess, Mike Hotek, Javier Loria, Adam Machanic, Antonio Soto, Adolfo Wiernik: *SQL Server 2005 Implementation and Management,* Microsoft Press, 2006.

Davidson, Cathy: *Now You See It: How the Brain Science of Attention Will Transform the Way We Live, Work, and Learn*, Viking, New York, 2011.

Devlin, Keith: *Sets, functions, and logic: an introduction to abstract mathematics*, Chapman & Hall/CRC mathematics, Boca Raton, Fla: Chapman & Hall/CRC, 2004.

Dieudonné, Jean: *Foundations of Modern Analysis* 1. Academic Press. NY, 1969.

Donald Knuth: *The Art of Computer Programming*, Addison–Wesley, NY, 1986.

Downing, Douglas: *Algebra the Easy Way*, Barron's Educational Series, Barron's, Hauppauge, N.Y., 2003.

Esnault, Hélène; Viehweg, Eckart: *Lectures on Vanishing Theorems*, Birkhäuser Verlag, Basel, Boston, 1992.

Eves, Howard Whitley: *An Introduction to the History of Mathematics*, Saunders, Philadelphia, 1992.

Garrison, Mark J.: *A Measure of Failure: The Political Origins of Standardized Testing*, State University of New York, Albany, 2009.

Harris,Smith and Harris: *The Myths of Standardized Tests: Why They Don't Tell You What You Think They Do*, Rowman & Littlefield, 2011.

Huddleston, Mark W. Boyer, William W.: *The higher civil service in the United States: quest for reform*, University of Pittsburgh Press, 1996.

Kate, S.K.; Bhapkar, H.R.: *Basics of Mathematics*, Technical Publications, Pune, 2009.

Lang, Serge: *Undergraduate Analysis*, Springer-Verlag, Berlin, New York, 1997.

Miller, Charles D.; Heeren, Vern E.: *Mathematical Ideas,* Scott, Foresman, NY, 1986.

Phelps, Richard P.: *Correcting Fallacies about Educational and Psychological Testing,* American Psychological Association, Washington, DC, 2008.

Phelps, Richard P.: *Standardized Testing Primer*, Peter Lang, New York, 2007.

Ricciardi, Luigi M.: *Lectures in Applied Mathematics and Informatics*, Manchester University Press, Manchester, 1990.

Roy S. Freedman: *Introduction to Financial Technology*, Academic Press, Amsterdam, 2006.

Shirali, Shailesh: *A Primer on Logarithms*, Universities Press, Hyderabad, 2002.

Sir Thomas Little Heath: *The Thirteen Books of Euclid's Elements,* Cambridge Univ. Press, NY, 1908.

Stewart, James: *Single Variable Calculus: Early Transcendentals*, Thomson Brooks/Cole, Belmont, 2007.

Stifelio, Michaele: *Arithmetica Integra*, Iohan Petreium, Nuremberg, 1544.

Van der Lubbe, Jan C. A.: *Information Theory*, Cambridge University Press, NY, 1997.

Vivian Shaw Groza and Susanne M. Shelley: *Precalculus mathematics*, Holt, Rinehart and Winston, New York, 1972.

Wegener, Ingo: *Complexity Theory: Exploring the Limits of Efficient Algorithms*, Springer-Verlag, Berlin, New York, 2005.

Zwillinger, Daniel; Kokoska, Stephen: *CRC Standard Probability and Statistics Tables and Formulae*, CRC Press, NY, 2010.

Index

❑❑❑